Pests and Diseases in Flower, Medicinal and Aromatic Crops

Pests and Diseases in Flower, Medicinal and Aromatic Crops

Dr. S. Parthasarathy
Assistant Professor (Plant Pathology)
Amrita School of Agricultural Sciences
Amrita Vishwa Vidyapeetham, Coimbatore

Dr. R. Kalaivanan
Assistant Professor (Plant Pathology)
Vanavarayar Institute of Agriculture, Pollachi
(Affiliated to Tamil Nadu Agricultural University)
Coimbatore

Dr. R. Sangavi
Assistant Professor (Entomology)
Vanavarayar Institute of Agriculture, Pollachi
(Affiliated to Tamil Nadu Agricultural University)
Coimbatore

Dr. P. Lakshmidevi
Teaching Assistant (Plant Pathology)
Institute of Agriculture, Kumulur, Trichy
Tamil Nadu Agricultural University, Coimbatore

-EPH-

Elite Publishing House

CRC Press
Taylor & Francis Group
Boca Raton London New York

CRC Press is an imprint of the
Taylor & Francis Group, an **informa** business

First published 2025
by CRC Press
4 Park Square, Milton Park, Abingdon, Oxon, OX14 4RN

and by CRC Press
2385 NW Executive Center Drive, Suite 320, Boca Raton FL 33431

British Library Cataloguing-in-Publication Data
A catalogue record for this book is available from the British Library

Print edition not for sale in South Asia (India, Sri Lanka, Nepal, Bangladesh, Pakistan or Bhutan).

ISBN13: 9781041034315 (hbk)
ISBN13: 9781041034346 (pbk)
ISBN13: 9781003623793 (ebk)

DOI: 10.1201/9781003623779

Typeset in A Caslon Pro
by Elite Publishing House, Delhi

-EPH-

Contents

About the Authors

Dr. S. Parthasarathy currently serves as an Assistant Professor in Plant Pathology at the Department of Plant Pathology at Amrita School of Agricultural Sciences. He earned his PG and Ph.D. degrees from Tamil Nadu Agricultural University, Coimbatore. Dr. Parthasarathy has been honored with various prestigious fellowships, including the BIRAC Post Master's Innovation Fellow in DBT (2015–17), the Canadian International Development Research Centre-Nano Project JRF (2013–14), the UGC Non-Special Assistance Programme (2012), and the Ministry of Human Resource Development Fellowship during his undergraduate studies in 2009. He has also received medals during his undergraduate program and numerous meritorious awards at both international and national conferences for his research contributions. An emerging expert in Plant Pathology, he has authored 55 research articles in peer-reviewed journals, written 16 books, and contributed approximately 45 chapters to various books of national and international acclaim. Furthermore, he has served as a member of the editorial boards of national journals and various international scientific societies. His areas of specialization encompass Biological Control and Molecular Plant Pathology.

Dr. R. Kalaivanan holds the position of Assistant Professor specializing in Plant Pathology at the Department of Crop Protection at Vanavarayar Institute of Agriculture in Pollachi. He obtained his postgraduate and Ph.D. degrees from the Agricultural College & Research Institute in Madurai (TNAU). As a young and accomplished Plant Pathologist, he boasts an impressive record, including the publication of 2 international research papers, 5 national research papers, a book, and contributions to 4 book chapters. He actively participates in several scientific societies and has a history of engagement in national and international conferences, earning multiple awards. Moreover, he is actively involved in extension activities, particularly in providing hands-on training to farmers to address real-world issues encountered in the field. His areas of expertise encompass Biological Control, Plant Bacteriology, and Molecular Plant Pathology.

Dr. R. Sangavi holds the position of Assistant Professor in Entomology at the Department of Crop Protection at Vanavarayar Institute of Agriculture, Pollachi. She completed her Ph.D. from Navsari Agricultural University, Gujarat. She is a recipient of the ICAR SRF fellowship and has authored 12 research papers, 5 book chapters, and 2 books. She also serves as a reviewer for two national journals and has organized several national and international conferences. Dr. Sangavi has represented as a resource person in various farmer training programs and has served as a research guide for students and a member of numerous agricultural projects. Her area of specialization includes Biological control, Biopesticides, and Acarology.

Dr. P. Lakshmidevi holds the position of Teaching Assistant in Plant Pathology at the Institute of Agriculture, Tamil Nadu Agricultural University. She completed her Ph.D. at Tamil Nadu Agricultural University, Coimbatore. She is a recipient of RGNF fellowships during 2011-14, a Young Plant Pathologist Award, and the Best Book Contributor Award in 2022. Dr. Lakshmidevi has conducted 2 Farm Testing trials, 4 Front Line Demonstrations, two farmers' field schools, and 43 on and off-campus training sessions. She has published 4 books with ISBNs, 12 research articles, 29 conference papers, and 23 book chapters with ISBNs.

1 Introduction

Horticulture deals with cultivation of garden crops, fruits and vegetables, while Floriculture focuses on cultivation of flowers and ornamental plants. India, being a land bestowed with favourable climate for agriculture, is conducive towards cultivation of flowers, fruits, vegetables and crops throughout the year.

Floriculture is an age-old occupation in India; however, it was after economic liberalisation that the Indian Government identified floriculture as a sunrise industry, rendering it 100% export-oriented status. Flowers such as rose, carnation, chrysanthemum, gerbera, gladiolus, orchids, anthurium, tulip and lilies account for the major share of exports. Flowers are widely used for religious purposes, decorations, gifting, medicines and extraction of perfumes and oils. While exports and religious festivals are key growth drivers for floriculture, demand for flowers is also increasing from metro cities owing to rising westernisation trends. Apart from decorative and gifting purposes, flowers are also used in industrial applications to manufacture flavours, fragrances, natural colours, medicines and Ayurveda. Owing to steady increase in demand of flower floriculture has become one of the important Commercial trades in Agriculture. Hence commercial floriculture has emerged as hi-tech activity-taking place under controlled climatic conditions inside greenhouse. Floriculture products mainly consist of cut flowers, pot plants, cut foliage, seeds bulbs, tubers, rooted cuttings and dried flowers or leaves. The important floricultural crops in the international cut flower trade are rose, carnation, chrysanthemum, garbera, gladiolus, gypsophila, orchids, anthurium, tulip, and lilies.

Present status and growing trade is still in infancy. Floriculture in India, is being viewed as a high growth Industry. Commercial floriculture is becoming important from the export angle. The liberalization of industrial and trade policies paved the way for development of export oriented production of cut flowers. The new seed policy had already made it feasible to import planting material of international varieties.

The government of India offers tax benefits to new export oriented floriculture companies in the form of income-tax holidays and exemption from certain import duties. Agricultural and Processed Food Products Export Development Authority (APEDA), responsible for export promotion and development of floriculture in India, grants subsidies for establishing cold storage, precooling units, refrigerated vans and green houses, and air freight subsidy to exports. It has been found that commercial floriculture has higher potential per unit area than most of the field crops and is therefore a lucrative business.

According to a report of the APEDA, the total area under flower crops was estimated around 34,000 hectares, which included 24,000 hectares under traditional flowers such as marigold, jasmine, aster, rose, chrysanthemum, tuberose and 10,000 hectares under modern flowers like carnation, rose, gerbera, gladiolous, anthurium. Returns from floricultural products were estimated at Rs. 205 Crores, which included Rs.105 Crores from traditional and Rs. 100 Crores from modern flowers. This phenomenal growth would not have been possible without the proactive policies implemented from time to time. Central sector schemes implemented by the National Mission for Integrated Development of Horticulture, Micro Irrigation, National Horticultural Board, Agriculture and Processed Food Export Development Authority played a pivotal role in achieving the growth.

Flower crops production has evolved considerably through new technological advances in irrigation, media and environmental control, along with the appearance of new centers of large scale production of plant material. These changes have necessitated the development of newer and innovative ways of suppressing pests and pathogenic fungi, bacteria, viruses, and nematodes. Insect pests are common problems in greenhouse and field production systems associated with floriculture crops. Many insect pests cause direct feeding damage to crops; however, several cause indirect damage by vectoring pathogens such as fungi and viruses. The primary insect pests of floricultural crops that vector or transmit diseases are the western flower thrips, fungus gnats, shore flies, green peach aphid, and sweetpotato whitefly. The pest and disease management strategies used are generally associated with agro-chemicals applications although extensive use of biocides may lead to resistance. Therefore, in addition to the use of biocides, and in order to effectively deal with insect vector populations, consideration should be given to cultural and sanitation practices and using biological control agents (natural enemies). However, these strategies must be effective in sustaining populations at very low levels in order to reduce the potential of pest and disease transmission to floricultural crops.

Medicinal and aromatic plants (MAPs) hold an important place in modern civilization owing to their roles in different kinds of industries. However, because of the increasing demand for raw material, medicinal plant growers are extensively using chemicals to fulfill their needs. The Medicinal and Aromatic Plants (MAPs) value chain is complex, with different industries involved in primary- and secondary-processed products and services. However, MAPs are facing the danger of extinction due to the attack of some severe insect pests, diseases, weeds, deforestation, extensive exploitation and harvesting from natural sources, and lack of proper knowledge on those problems among the majority of the people. Therefore, there is an urgent need for a thorough investigation of the cultivation management of MAPs. Reports on insect pests of MAPs are meager and scattered. About 40-45 major destructive insect-pest species are found, causing a loss of about 50-60% damage to MAPs. Among these factors, insects and disease are the most important factors for causing significant damage to plants. Pesticides are used to manage harmful pests and diseases to enhance crop yields.

Additionally, these synthetic chemicals might alter the therapeutic values of herbal products. Indiscriminate usage of pesticides causes severe ill effects on public health and the environment. Research on pests and diseases helps understand the life history of pest species to formulate suitable management strategies. Moreover, eco-friendly management practices are essential. Monitoring of pest populations and disease epidemiology, identification, and choosing a combination of methods to keep pests and diseases at an acceptable level is necessary. The present chapter deals with overviews of the diversity of pests and diseases of MAPs and integrated pest management strategies for dealing with them in an eco-friendly manner.

References

Trenbath, B. R. (1993). Intercropping for the management of pests and diseases. Field crops research, 34(3-4), 381-405.

Alford, D. V. (Ed.). (2008). Pest and disease management handbook. John Wiley & Sons.

Scopes, N. E. A., & Ledieu, M. S. (1980). Integrated pest control on chrysanthemums and other flower crops. Scientific Horticulture, 31, 48-53.

Bout, A., Boll, R., Mailleret, L., & Poncet, C. (2010). Realistic global scouting for pests and diseases on cut rose crops. Journal of economic entomology, 103(6), 2242-2248.

Gahukar, R. T. (2018). Management of pests and diseases of important tropical/ subtropical medicinal and aromatic plants: A review. Journal of Applied Research on Medicinal and Aromatic Plants, 9, 1-18.

Shivakumara, K. T., Roy, S., & Meena, R. P. (2023). Integrated Pest and Disease Management in Medicinal and Aromatic Plants-Based Cropping System. In Integrated Pest Management in Diverse Cropping Systems (pp. 517-548). Apple Academic Press.

2 Anthuriums

INSECT PESTS

Whitefly, *Aleurotulus anthuricola*

Damage symptoms

- Adults and juveniles with white waxy coats live under the leaf surface and in the petiole sheath.
- Infested plant parts will turn yellow, wilt, and dry as the insect feeds on the plant sap.

Pest identification

- Whitefly larvae look like tiny scales and live on the undersides of affected leaves.

Management

- Pick up the infected leaves and burn them.
- Place yellow sticky traps across the neighbourhood to ensnare and kill adults.
- Use systemic pesticides such as Methomyl 1.8 L or Dimethoate 30 EC to spray for insects (reduces eggs, nymphs, pupae and adults).
- Adults and developing larvae can be killed with 1 ml/L of dichlorvos 76 EC or 50 EC malathion.
- The effectiveness of Bifenthrin 10 WP was restricted to adults.
- The use of Acephate 75 SP led to a decrease in the population of both nymphs and pupae.

» Parasitic wasp *Encarsia formosa* was released into the wild at a density of three insects per square metre.

Thrips, *Chaetanaphorothrips orchidii*

Damage symptoms

» First appearing in March and peaking in August/September, damage occurs year round.

» Adults and juveniles alike feed on leaf sap.

» Curled or bent leaves that have been infested will often have silvery grey scars or calloused areas where feeding has occurred.

» As they feed on the cell sap of their hosts, thrips cause white or brown streaks and spathe deformation.

» Observable mottling of plants and blossoms.

» Tomato spotted wilt virus is spread by thrips, which can infect a wide variety of houseplants.

Pest identification

» Thrips are very skinny insects that are very little (less than 1/20 inch).

» The wings of adult thrips have a distinctive long fringe of hairs down their edges.

Management

» Spraying the foliage with a combination of Cyfluthrin 2.0 EC and Chlorpyriphos at 1ml/L helps reduce the amount of damaged blossoms.

» The thrips population can be reduced by spraying either 1 ml/L of Dimethoate 30 EC or 0.5 ml/L of Phosphamidon 85 WSC.

» Apply either 25 EC of oxademeton methyl, 30 EC of dimethoate, 1 mL of profenofos, or 1.5 g of acephate per litre of water.

» After 15 days, spray the area with 10 ml/L of Pongamia or Jatropha oil.

» Depending on the severity of the outbreak, spray Imidacloprid 200 SL at 0.4 ml/L or Fipronil 5 SC at 1.5 ml/L.

» Amblyseius cucumeris, a predatory mite, was released into the field at a density of 500,000 mites per hectare.

Aphid, *Myzus circumflexus*

Damage symptoms

- Adults and juveniles alike feed on the plant's sap.
- Symptoms include stunted growth, yellowing leaves, and leaf deformation.
- Whenever honeydew is secreted, black sooty mould begins to flourish.
- Aphids can distort new growth or, in severe circumstances, stunt plants.

Management

- Dicofol, Malathion, or Phosphomidon 2 ml/L dilution spraying.
- The extract of the pyrethrum plant proved to be an effective ingredient in a mixture used for pest management.
- Predator *Adalia bipunctata* is effective.
- Parasitic wasp *Aphidius colemani*, released at a rate of 5/ m^2 per week on the field.

Red spider mite, *Tetranychus cinnabarinus*

Damage symptoms

- Mite colonies are visible on bottom surface of mature leaves.
- They feed on the plant›s sap, resulting in spots, yellowing, and ultimately the wilting of the damaged leaves.
- Both the upper and lower sides of the leaf will turn a bronzy colour.
- The leaves wither and turn brown.
- When mite numbers rise, they can cause webbing, leaf drop, and ultimately plant death.

Pest identification

- Males and females, nymphs and adults, all have a common red colouring.
- A white, round egg is placed on the underside of the leaves.

Management

- To control pests, spray Dicofol 18.5 EC at 2.5 ml/L, Triazophos 40 EC at 1.5 ml/L, or Ethion 50 EC at 1 ml/L, then apply Pongamia or Neem oil at 10 ml/L 15 days later.

» Abamectin 1.9 EC at 0.5 ml/L or Flufenoxuron 10 DC at 1 ml/L should be sprayed if the infestation is particularly bad.

» At least two sprays of Kelthane (1 ml/L) or Vertimec (0.4 ml/L) should be applied first thing in the morning.

» Predator mite *Phytoseiulus persimilis* was released at a rate of 6/ m^2 in the field.

DISEASES

Rhizoctonia root rot, *Rhizoctonia solani*

Symptoms

» Whenever a plant's young, vulnerable stems are girdled, they quickly swell and collapse under the weight of the plant.

» Plant roots and lower stems are particularly vulnerable to rhizoctonia infection.

» It can also invade and propagate in the top leaf canopy during moist periods.

Favorable conditions

» Soils that have been saturated with water are ideal for the spread of disease.

Survival and spread

» Without a host plant, rhizoctonia can live in soil for years.

» The fungus grows in mats of sclerotia, which are made up of mycelia that are braided together tightly.

» The fungus relies on sclerotia, which are similar to seeds, to endure harsh environments.

» One of the ways the fungus spreads in nurseries is via the small sclerotia that adhere to the surfaces of trays and pots.

» Infected cuttings or seedlings are a common entry point for *Rhizoctonia* into manufacturing operations.

Management

» Never use untreated native soils in media blends.

» Choose potting soils with good drainage.

» Never keep potting media mixes, peat moss, sphagnum moss, or chips on bare soil, as these materials are susceptible to colonisation by fungi.

- Growing plants on raised benches reduces the amount of soil that the roots have to absorb.
- Fungicides like thiophanate methyl, fludioxonil, and flutolanil can be used to control Rhizoctonia infections.

Root rot, *Phytophthora nicotianae* var. *parasitica*, *Pythium splendens*

Symptoms

- Roots are a primary target for *Phytophthora* and *Pythium*.
- The plants will still wilt, even though there is plenty of water in the soil.
- The sloughing off of roots is the basic method of diagnosis.
- In extreme weather, the leaves may develop black or brown lesions.
- Infected plants wilt, their leaves become yellow (chlorosis), and their roots dry out.

Favorable conditions

- Soils that have been saturated with water are ideal for the spread of disease.
- Use of light, well-drained soil mixtures is a tried-and-true method for warding off these afflictions.

Management

- Make sure to use synthetic soil mixtures with enough drainage.
- Use only disease-free stock plants.
- Diseased plants should be thrown away, and a fungicide drench applied to the rest of the productive area.
- To prevent disease transmission, pots that have previously held soil should be thoroughly cleaned and disinfected before being used again.
- When switching plants, make sure to clean shears, knife, and other instruments.
- Mefenoxam (Subdue, Maxx), Fosetyl-AL (Aliette WDG), Dimethomorph (Stature), Fluopicolide (Adorn), and Phosphorous acid (Alude, K-Phite) are all examples of fungicides that could be utilised.

Anthracnose/Black nose disease, *Colletotrichum gloeosporioides*

Symptoms

» Small brown to black specks on the spadix of the flower are the first sign of trouble (nose).

» In a short amount of time, these patches can balloon to balloon-like proportions, become fluid, darken from brown to black, and perhaps engulf the entire spadix.

» At some point, the spadix could become detached.

» Acervuli, or black formations that hold spores, can also be seen by growers on decaying leaves and stems.

» Selling flowers or plants in flowering containers is prohibited.

Favorable conditions

» Wet and warm weather is ideal for the spread of this disease.

» Plant tissues that have been injured by herbicides, fertilisers, or bacterial blight are easy prey for *Colletotrichum* (*Xanthomonas*).

Survival and spread

» Spores produced by fungi are extremely minute and dispersed easily by splashing water, wind or human hands.

Management

» This disease can easily spread throughout a manufacturing facility without a rigorous cleanliness regimen.

» Effective fungicides include Mancozeb (Dithane M-45) and Benomyl (Benlate). Chemical residues reduce the plant and flower›s resale value, hence fungicide treatment is generally discouraged.

» Disease resistance has been bred into several newer cultivars of Anthurium thanks to efforts in both Hawaii and Florida. Fortunately, black nose is quite uncommon in modern cultivars because of their inherent resistance to this virus.

Bacterial blight, *Xanthomonas campestris* pv. *diffenbachiae*

Symptoms

» Yellowed (chlorotic), water-soaked lesions along the leaf margins are the first obvious symptoms and they spread swiftly.

- Anthurium leaves develop V-shaped, water-soaked necrotic sores along the margins.
- The bacterial invaders swiftly colonised the entire plant.
- The leaves of plants with systemic infections may take on a bronze hue.
- There is a risk that flower quality will decline and/or blossoms will become unsalable.
- The life cycle of plants always ends in death.

Favorable conditions

- Guttation droplets arise in conditions of high temperature, high humidity, and saturated soils.
- Wet surfaces provide a swimming pool for bacteria.

Survival and spread

- Transmitting disease by cuttings or division is a major risk while growing plants.
- Harvesting flowers or chopping down ageing bushes can potentially spread disease.

Management

- Plants grown in sterile tissue culture.
- Bacteria can swim across damp surfaces, thus it is crucial to keep the plant life dry.
- Drip irrigation is the best method for achieving this result.
- Disease rates can be lowered through the use of effective cultural practises and rigorous phytosanitation.
- Reduce the greenhouses humidity and temperature by increasing ventilation. Leave some room on the bench for the plants.
- Plants that show signs of infection should be thrown out promptly.
- Copper-based products (Kocide 3000 O, Phyton 27 T/O, Dithane) and those containing *Bacillus subtilis* (Cease) have proven to be effective.

Bacterial wilt, *Ralstonia solanacearum*

Symptoms

» The chlorosis, or yellowing, of leaves is often the first sign noticed.

» Veins in the plant's leaves and stems turn a brownish bronze as the disease quickly spreads through the vascular system.

» If the stems of highly infected plants are severed, a brown slime-like substance will flow out.

» It doesn't matter how much water is in the soil, plants will still show signs of wilting.

Favorable conditions

» There is a risk that symptoms could be temporarily hidden and bacteria could spread more easily in the cool conditions of a greenhouse.

» In hot heat, symptoms manifest swiftly.

Survival and spread

» Soil is an ideal environment for bacteria to live in.

» Wilt caused by bacteria can spread through contact with infected soil, water, tools, or workers.

» There is a high risk of transmitting the bacterium that causes wilt by using cuttings that have been infected with the disease.

Management

» The most effective method of controlling the spread of this infection is a rigorous cleanliness plan.

» Use only disease-free seeds and cuttings.

» Both the plants and the soil they were growing in should be thrown away if they were tainted.

» The use of phosphoric acid-based fungicides has also been demonstrated to be effective.

References

Hata, T. Y., & Hara, A. H. (1992). Anthurium thrips, Chaetanaphothrips orchidii (Moulton): biology and insecticidal control on Hawaiian anthuriums. International Journal of Pest Management, 38(3), 230-233.

Hara, A. H., Nishijima, W. T., Hansen, J. D., Bushe, B. C., & Hata, T. Y. (1990). Reduced pesticide use in an IPM program for anthuriums. Journal of economic entomology, 83(4), 1531-1534.

Hara, A. H., Nishijima, W. T., Hansen, J. D., Bushe, B. C., & Hata, T. Y. (1990). Reduced pesticide use in an IPM program for anthuriums. Journal of economic entomology, 83(4), 1531-1534.

3 Carnation

INSECT PESTS

Thrips, *Frankliniella schultzei*

Damage symptoms

» The nymphs and adults alike feed on the sap of plants.

» Buds with extensive damage will have brown streaks and become deformed with scorched borders, lowering quality.

» Their faeces are initially brown, but eventually turn black.

» In extreme cases of infection, the leaves may wither and turn silvery.

» Damage to developing buds prevents them from opening, resulting in a significant drop in flower production.

Pest identification

» Female thrips reach maturity with a slender, 1-2 mm long body that is pale brown or grey. The males of this species are the ones that are smaller and yellow. Both sexes have four slender, hairy wings.

Favorable conditions

» The thrips population responds mostly to temperature changes.

Management

» Both Dimethoate 30 EC and Oxydemeton methyl 25 EC should be sprayed at 2 ml/L at 15-day intervals.

» To lessen the number of pupae in the soil, inject a solution of Chlorpyriphos

20 EC at a concentration of 4 ml/L.

- Use 0.4 ml/L of Imidacloprid 200 SL, 1 ml/L of Fipronil 5 SC, or 1.5 g/L of Acephate 75 SP to spray the area.
- Apply a spray containing either 1 g of Asatap (Acephate), 0.5 ml of Decis (Deltamethrin), or 1 g of Exodust.

Bud borer, *Helicoverpa armigera*

Damage symptoms

- In September and October, females lay single eggs on developing buds, causing damage that lasts all the way through February and March.
- The larvae tunnel into the growing petals of a bud and eat them, leaving the petal hollow.
- The larvae also eat the petals off of open blooms, which greatly reduces the harvest.
- Negative effects on plant development are seen.
- They drill distinctive circular holes in flower buds and blossoms.

Pest identification

- In terms of colour, eggs are often a pale cream.
- When fully developed, caterpillars reach a length of around 40 mm and display a wide range of colours, from yellow to green to pink to brown, as well as dark flecking and longitudinal pale and dark stripes.

Favorable conditions

- The polyhouse has a warmer environment.
- There was a change in the relative humidity in the polyhouse.
- Changes in polyhouse temperature.

Management

- Capturing and killing off the adult larvae.
- Spraying at 15-day intervals with 25 EC Quinalphos, 1 EC Methyl parathion, or 35 EC Endosulfan.
- If the outbreak is really bad, spray Indoxacarb 14.5 SC at 1 ml/L or Thiodicarb 75 WP at 1 g/L.

- Neem formulations at 1-2 ml/L or Neem seed kernel extract (40 g/L) should be applied after HaNPV at 250 LE/ha has been applied.
- Use 2 g/L of *Bacillus thuringiensis*
- Let loose 25,000,000 egg-parasitic Trichogramma/ha.
- Introduce 10 N/ha of Helilure sex pheromone traps.
- One can use 0.75 ml/L of Spinosad or 0.5 ml/L of Thiodicarp in a spray.

Aphids, *Myzus persicae*

Damage symptoms

- These parasitic bugs are responsible for a lot of human suffering.
- Buds and the undersides of leaves are typical feeding targets.
- Insects that gorged themselves on tender young leaves grew into deformed adults.
- Sometimes, chlorotic dots will appear in older leaves.
- Colonies of nymphs and adults feed on the sap of leaves, stems, and flower buds.
- Honeydew secreted by aphids encourages the growth of black fungus on plants.
- The mottle virus can also be spread by this animal.

Favorable conditions

- It is easier for aphid populations to rapidly expand when the sky is overcast.
- Aphid juvenile development is aided by cool temperatures

Management

- Before the flowers open, spray them with Dimethoate 30 EC at 1 ml/L.
- Apply Thiomethoxam at 1 ml/L, Imidachloprid at 2 ml/L, or Asatap (Acephate) at 0.5 to 1.0 g/L of water to the plants.

Tortrix moth, *Tortrix pronubata*

Damage symptoms

- The caterpillar spins a silken web around the leaves and enters it to find shelter and food.

- Damage typically observed at or near the tips of developing shoots.
- It is possible for the larvae to burrow into the flower buds as well as eat the leaves.

Management

- For effective treatment of tortrix moth larvae in nurseries and beds, a combination of Carbaryl and Dicofol should be used.

Red spider mite, *Tetranychus urticae*

Damage symptoms

- This bug causes significant damage to greenhouse-grown carnations, especially between the months of September and November, and between the months of February and May.
- On the underside of the leaves, beneath the filthy webs, are colonies of mites.
- Leaves become brown and shrivel up when affected.
- Buds, blooms, and eventually the entire plant can get infested.
- They cause tiny yellow or white spots on the leaves by sucking the plant's sap.
- When a plant's leaves turn bronze, they generally die off too soon.
- Plants that are attacked lose their strength, slow down in growth, and, in the worst situations, die.

Pest identification

- Each mature female is little about 0.5 mm in length. The average length of a mature male is 0.3 mm. They have a yellowish green colouring and are marked by two dark spots, one on each side of the body. Males have smaller, more tapered bodies with fewer and less noticeable marks. The two reddish eye-spots are shared by both sexes.

Favorable conditions

- Intense heat and dry air.
- Larger harvest (dense crop).

Management

- Disposal of infected plant material via cutting and burning.
- A heavy misting of water on the plant will wash away the webs.

- At 15-day intervals, spray Wettable Sulfur 80 WP at 3 g/L, Profenofos 50 EC at 1 ml/L, or Ethion 50 EC at 1 ml/L, and then apply 10 ml/L of Pongamia or Neem oil.
- In the summer, when infestations are at their worst, spray a combination of Abamectin 1.9 EC at 0.5 ml/L and Flufenoxuron 10 DC at 1 ml/L or Fenazaquine 10 EC at 1 ml/L.
- *Amblyseius tetranychivorus* is a predatory mite that should be released at a rate of 20 mites per plant.
- It was reported that Cultivars Isaq, Alcardi, and Sunrise were resistant to mites.

DISEASES

Fusarium wilt, *Fusarium oxysporum* f. sp. *dianthi*

Symptoms

- The leaves will turn yellow and the plant will produce crooked or bowed shoots first.
- The greenery shrivels and turns brown.
- Initially, the symptoms only appear on a small number of branches or a localised area of the plant, but they will spread until the entire plant is affected.
- When stems are sliced open, the vascular area reveals a brown tint.
- Even the pith and cortex might get discoloured at times.
- The interior tissue of the stems is shredded at the end.
- Symptoms of wilt appear on mature plants over the course of several months before they die and turn straw colour.

Favorable conditions

- The temperature is a major factor in the spread of the disease, soil temperatures of 24°C or above are ideal.
- Infection requires damp circumstances for at least 8 hours, and preferably 10 hours.

Survival and spread

- This pathogen can be found in the soil and can live for a long time in decaying plant matter.

- Irrigation water and airborne spores both contribute to the spread of the fungus.
- Infected seeds and soil propagate the disease.
- Root-to-root contact and the transport of tainted soil by hands and tools were major vectors of infection.

Management

- It's important to remove and burn the infected plants.
- 30 days of soil solarization using 0.1 mm thick clear polyethylene film.
- Kavach (Chlorothalonil) or Alleitte (a pyrethroid) sprayed at 1.5 g/L is equivalent to the same concentration in water.
- Carbendazim 0.1%, difenoconazole 0.05%, or *Bacillus amyloliquefaciens* 0.5% drenched into the soil once a month is recommended.
- *Pseudomonas fluorescens* applied at 25 g/m^2 to the soil and 0.5% to the leaves, once a month.
- Disease was reduced even when vulnerable cultivars like Alice, Fulvio Rosa, Gus Royalette, and Johy were grafted onto resistant rootstocks like Arancio 25D, Exquisite, Heidi, and May Britt and grown in naturally contaminated fungus-rich soil.
- *Trichoderma harzianum* should be applied to soil just after it has been treated with steam or chemically disinfested.

Black spot, *Alternaria dianthi*

Symptoms

- The earliest sign is the appearance of tiny purple dots on the leaves.
- As more and more spores begin to form, they quickly grow in size, with the core eventually turning a dark brown and eventually black.
- Infected knots on a stem typically manifest the disease.
- Yellowing of the adjacent leaf tissue is observed here.
- The leaves quickly wilt and die due to the severe disease.
- Flowers are not safe from the fungus's infection, either.

Management

- Having enough air flow across a space might help keep humidity levels down.
- It's important to use plant material that hasn't been exposed to any diseases.
- Chlorothalonil spray at 1.5 g/L concentration.

Grey mold, *Botrytis cinerea*

Symptoms

- When flowers rot, they release fuzzy grey fungal spores that can infect other parts of the plant they touch.
- Fuzzy, greyish spores that form on the surface of the rotting tissues.
- Affected tissues become dark and rotten due to the fungus.
- Blossoms are the most negatively impacted, leading to subpar quality.

Favorable conditions

- Circumstances of low temperature, high humidity, and high moisture content are ideal.

Survival and spread

- The pathogen is still alive in the dead plant.
- Disease spores can travel great distances in the air.

Management

- Snip off spent blooms and throw them away.
- Continue the flow of air horizontally.
- Reducing humidity is a priority.
- Iprodione and Fenhexamid help mist blossom.
- By applying Tebuconazole 250 EC at 0.15% and Difenoconazole 25% EC at 0.15% as a foliar spray, we were able to decrease the occurrence of blossom blight.

Rust, *Uromyces dianthi*

Symptoms

- Even though the disease is most noticeable on the leaves, it can be seen on other green plant components.

- At first, a few pale green dots will show up. Due to the proliferation of uredospores, they transform into powdery brown blisters.
- Plant organs that have been severely infected are distorted.

Favorable conditions

- Thriving in wet environments.

Survival and spread

- Only plant life can provide a means of transport.
- Uredospores spread from sick plants to healthy crops through the air.
- Spread via air currents or splashes of water.

Management

- Make sure the plants use to take cuttings are disease-free.
- Make sure there is adequate ventilation in greenhouses to avoid the formation of moist or humid areas, which can promote the spread of disease. Spread plants out and airflow.
- When watering, make sure the leaves do not get soaked.
- Take away and burn the contaminated leaves.
- Before working with healthy plants, make sure hands and tools are clean.
- Protecting from infection with the fungicide Mancozeb and using it every two weeks as soon as the first symptoms will help stop the disease in its tracks.
- When plants are attacked, using a fertiliser high in potash can aid in their recovery.
- Fungicides made with myclobutanil have shown promise in controlling infestations.
- Penconazole, Moncozeb (also used for prophylaxis), and Triticonazole are alternatives to Myclobutanil.

Foot rot, *Phytophthora nicotianae* var. *parasitica*

Symptoms

- If it rains too much, the leaves and flower buds will get sick.
- Symptoms of infection include wilting, browning of the collar, and ultimately death in infected plants.

- Leaves turn yellow and die, the stems brown on the outside and the nodes on the inside.
- Symptoms include root rot and eventual death for infested plants.
- Drying causes a shortening and blackening of the stems, and eventually the stems decay and turn purple.
- Dying and discoloration of the leaves occur from the underside up when the roots are afflicted.
- The most noticeable symptom is the soft decay of young leaves and floral buds.

Favorable conditions

- Viruses and bacteria both do well in humid environment
- This fungus may thrive at temperatures between 10 and 35 °C, while 27 °C is ideal.
- In contrast to a temperature range of 15-20 °C, disease spreads more rapidly between 25 and 35 °C.

Management

- Aleitte or Chlorothalonil at a concentration of 1.5 g/L in water should be used.
- Infections can be controlled with the help of fungicides like Metalaxyl + Mancozeb 0.2% and Fosetyl-AL 0.2%.

Stem rot, *Sclerotinia sclerotiorum*

Symptoms

- Stem rot causes slow, even wilting, unlike the sudden wilting caused by Fusarium wilt.
- The leaves turn a pale straw hue.
- A void has formed inside the affected stems.
- In a longitudinal cut of the diseased stem, brown sclerotia of various sizes and shapes are visible.
- Mycelia of *Sclerotinia sclerotiorum* colonise the root zone.
- Damaged plants develop a brown vascular system.

Favorable conditions

» The disease flourishes best in damp, cool soil.

Survival and spread

» Primary inoculum comes from contaminated soil and plants. The sclerotial bodies of the pathogen are what live in the soil.

Management

» Azoxystrobin, Tebuconazole + Trifloxystrobin, and Carbendazim are all excellent fungicides for saturating the soil to prevent stem rot.

» Biocontrol agents such as *Bacillus subtilis*, *Bacillus amyloliquefaciens*, and *Bacillus cereus* could be soaked into the soil.

» Application of *B. amyloliquefaciens* (VB7: 0.5% (5 ml/L) via root dip and soil soaking reduced the incidence of stem rot to 4.6%.

» *S. sclerotiorum* growth was inhibited by *Trichoderma asperellum* (NVTA2) and *T. harzianum* (NVTH1).

Root rot, *Pythium* spp.

Symptoms

» When a pathogen gets in at the collar, the leaves gradually turn brown and dry out from the bottom up.

» The plant's infection causes the stem to rot at the root, killing it.

» The plant's growth is impeded and eventually dies from the disease.

Favorable conditions

» Increased soil moisture, high relative humidity, and agricultural wastes all contribute to the growth of the disease.

Survival and spread

» This pathogen lives on in the soil and the remains of infected plants.

» Irrigation water is a major vector for its proliferation.

Management

» Apply a spray solution containing either 2.0 g/L Carbendazim (Bavistin), 1.5 g/L Aleitte, or 1.5 g/L Chlorothalonil (Kavach).

- Drenching the soil with Azoxystrobin at 0.15% or Difenaconazole at 0.05% aids in the control of the disease.
- The disease can be controlled by using Trichoderma viride at a rate of 2.5 kg/ha in 250 kg of the compost.

Bacterial wilt, *Burkholderia caryophylli*

Symptoms

- The tops of mature plants become droopy and pallid.
- The disease is characterised by the development of elongated discoloured stripes on the stem, as well as the development of splits in the stem.
- The cortical tissue has become sticky and discoloured, and the roots have decayed partially.
- There is a twist in the leaves.
- On average, one side of the roots is missing and the other side remains discoloured.
- There's a brown streak that runs up from the base of the cut.

Favorable conditions

- Increased by hot weather.

Survival and spread

- Foliage and litter.
- Water acts as a medium for the dissemination of bacteria.

Management

- Plants must be disease-free in order to be used.
- To propagate plants from cuttings, select those that originate from the higher regions of healthy stock plants.
- The remains of infected plants need to be gathered and burned.
- Watering from above or watering with splashes should be avoided.
- Some plant varieties are resistant to bacterial wilt, and they include Elegance, Northland, and Starlite.

References

Jawaharlal, M., Ganga, M., Padmadevi, K., Jegadeeswari, V., & Karthikeyan, S. (2009). A technical guide on carnation. Tamil Nadu Agricultural University, Coimbatore, 1-56.

Allwood, M. C. (1920). Carnations their Pests and Diseases. Journal of the Royal Horticultural Society, 45(2-3).

Khristova, E., & Aleksandrova, Í. (1960). Diseases and pests of Carnation and means of their control. Ovoshcharstvo, 7(7).

Mitrea, I., Mitrea, R., Tuca, O., & Stan, C. (2008, April). Integrated pest and disease management in greenhouse Dianthus caryophyllus L. production. In International Symposium on Strategies Towards Sustainability of Protected Cultivation in Mild Winter Climate 807 (pp. 751-756).

Reddy, P. P., & Reddy, P. P. (2016). Carnation. Sustainable Crop Protection under Protected Cultivation, 345-354.

4 China Aster

INSECT PESTS

Flower caterpillars, *Helicoverpa armigera* and *Phycita* sp.

Damage symptoms

- The larvae of *H. armigera* are known to eat the florets and petals of flowers, rendering them unmarketable.
- They cause a significant drop in yield by devouring flower buds' ovaries and stamens.
- Phycita sp. larvae eat the developing ovaries and filaments of the stamens, damaging the flowers in the process.
- Caterpillar faeces can be spotted on the ruined blossoms.
- Sadly, the flower has been misplaced.

Pest identification

- The caterpillar of *Phycita* sp. is green with a brown head.
- Greenish-colored pupae.
- Moths in their adult stages are grey.

Favorable conditions

- When temperatures drop, *Helicoverpa armigera* becomes more common.

Management

- Remove and dispose of infected flower buds and full-grown larvae.

- Flowers and buds can be protected from pests with a single spray of 0.01% Fenvalerate and a subsequent spray of 0.07% Endosulfan or 0.05% Methyl parathion.
- At the first sign of egg laying on young leaves, spray a mixture of Diflubenzuron 25 WP (2 g/L) and Endosulfan 30 EC (2 ml/L) or Methyl parathion (50 EC) (1ml/L) or Phenthoate (50 EC) (0.5 ml/L) or Fenvalerate (20 EC) (0.5 ml/L).
- Apply a 4% extract of Neem seed kernels.
- Don't forget to use 250 LE/ha of NPV spray on Ha.
- Dangerous fungus *Helicoverpa armigera* is infected with *Nomuraea rileyi.*

Stem borer, *Platyptilia molopias*

Damage symptoms

- White, elongated eggs are laid by adult moths on young, sensitive leaves and shoot tips.
- Caterpillars eat through the trunk and any lateral branches.
- The hollowing and wilting of shoots is a direct outcome of their nutrition.
- Black poop can be seen on the surface of damaged shoots.
- The affected growths wither and dry up.
- The attack from the pests could be fatal for the plants.

Pest identification

- White, elongated eggs.
- Butterflies and moths have a black head whereas caterpillars are cream-colored.
- Greenish-colored pupae.

Management

- The infected shoots are gathered and thrown away.
- When eggs are spotted, spraying with 0.005% Cypermethrin is effective, followed by 0.01% Ethofenprox, 0.05% Chlorpyriphos, 0.05% Phenthoate, or 0.05% Lambda-cyhalothrin.
- Application of Carbofuran 3 G granules at 1 kg a.i. /ha (33 kg/ha) and spraying with either Profenophos 50 EC at 1 ml/L, Endosulfan 35 EC at 2 ml/L, or Ethofenprox 10 EC at 1 ml/L.

Aphids, *Aphis gossypii*

Damage symptoms

» Aphids tend to cluster towards the bottom of flowers and at the base of petals.

» Stunted development, curling, yellowing, and eventual death of the China aster are all results of aphids sucking the plant's juices.

» Aphids tend to cluster around new growth, such as flower buds, leaves, and stems.

Spread

» Aphids are transferred from plant to plant by brown ants (*Lasius alienus*).

Management

» Apply either 25 EC/ 2 ml/L of oxydemeton methyl or 30 EC/ 2 ml/L of dimethoate as a spray.

» 1% Neem or Pongamia oil spray can be used in a pinch.

» Insecticidal soap (Neem or Pongamia, 1% strength) should be sprayed on the plants once a week to keep the insects at bay.

Leaf hopper, *Macrosteles quadrilineatus*

Damage symptoms

» The symptoms might range from a general yellowing of the foliage to the disappearance of veins in the leaves to the development of extra branches to the greening of the flower petals.

» The insect obtains its nourishment by sucking the plant juices from the green sections of plants using its piercing-sucking mouth parts; this causes the leaves to turn white and mottled.

» Leaves and stems of infected mature plants turn yellow (chlorotic), and the plants themselves become stunted and twisted.

» The flowers have lost their colour, become mottled, warped, and are infertile.

» Honeydew, produced by some species, can encourage the growth of sooty mould, a black fungus.

» Curling, yellowing, and even death of leaves are all symptoms of the damage.

» The disease severely disfigures plants, reducing both their aesthetic and commercial value.

» Cause the spread of the aster yellows virus.

Pest identification

» While adults are typically black or brown, nymphs can be either light or dark in appearance.

» The adults of the aster leafhopper species are around 4 mm in length and have an olive green, wedge-shaped body. Due to the three sets of spots on their heads, these insects are also known as six-spotted leaf hoppers.

Management

» Infested plants should be removed and disposed of to prevent further infestation.

» The spread of aster yellows can be slowed by eradicating the weed reservoirs and diseased crop plants that harbour the disease.

» Start using the yellow sticky traps in early April.

» To avoid being attacked by these insects, use a reflective mulch.

» When leaf hopper populations explode and the rain comes down often, chemicals with limited residual action, like synthetic pyrethroids, must be treated twice or thrice weekly.

Mealy bug, *Phenacoccus parvus*

Damage symptoms

» Extremely high infestation levels in the collar area as well as underneath plant components.

» Symptoms of infestation included reduced growth and the absence of blooms on the affected plants.

» Mealy bugs fed on the roots of the plants in the field, stunting their development even further.

Pest identification

» Body shape is that of an oval or elongated rectangle, flattened from dorsal to ventral.

» Color is a pale yellow, and the skin is covered in a fine, white wax powder.

» Legs are a pale yellowish brown; there are 17 to 18 tiny wax filaments around the female's body; the ovisac is elongated on the underside.

Spread

- » Moved to new locations by wind or water.
- » Crawlers are often carried by the wind for distances ranging from a few metres to a few kilometres.

Management

- » Sticky traps strategically placed around the field.
- » The most effective sprays are Sulfoxaflor, Abamectin + Thiamethoxam, and Spirotetramat.
- » Use 1% Neem or Pongamia insecticide soap and spray it around.
- » To let go the predator, Cryptolaemus montrouzieri.
- » Red spider mite, *Tetranychus urticae*

Damage symptoms

- » August and September are the months when the mite population peaks.
- » Colonies of mites, whose webs are covered in soil, can be seen on the undersides of leaves.
- » Nymphs and adults alike consume the sap for nutrition.
- » Foliage curls and the plant as a whole becomes yellow after mites have done their damage.
- » The leaves fall off and get a funny tint when this happens.
- » Stunted growth, wilting, and eventual death characterise plants that have been severely attacked.
- » They spin a delicate white web across the plant, much like a spider's web.

Pest identification

- » Mites tend to be tiny and a ruddy colour.

Management

- » Severely affected plant sections should be hacked off and burned.
- » Use of an insecticidal soap (1% Neem or Pongamia) once every seven days.
- » Apply 1% Pongamia oil after spraying Dicofol 18.5 EC at 2.5 ml/L, Wettable sulphur 80 WP at 3 g/L, Ethion 50 EC at 1 ml/L, or Profenofos 50 EC at 1 ml/L.

DISEASES

Fusarium wilt, *Fusarium oxysporum* f. sp. *callistephi*

Symptoms

- A vascular system-affecting soil-borne disease that ultimately kills the plant.
- Unnaturally slow development.
- Plants wilt and turn yellow, usually on one side, due to heat stress.
- The vascular system begins to turn a brownish tone.
- As a result, the plant's leaves curl, its colour fades, its roots decay, and eventually it dies.
- Even at soil temperatures of 75° to 80°F, disease can cause damping off in young seedlings.

Favorable conditions

- When soil and/or air temperatures are high, diseases are more likely to spread rapidly.

Survival and spread

- Typically spread through seeds.
- The plant can stay dormant in the ground for a long time.

Management

- Soil solarization.
- Seeds free of disease should be used.
- A 30-minute soak in 1% NaOCl for treating seeds.
- Once an aster has been contaminated, it is preferable to remove it and throw it away.
- The issue can be avoided if adequate drainage is provided for the soil.
- Clear any clutter from the planting area.
- Non-host crops should be rotated every 5 years.
- Avoid watering the sky.
- Putting in a raised bed for planting.
- Areas that have had Fusarium wilt in the past should not be replanted.

- Carbendazim 1 g/L soil drench.
- Try using 'Gem mixed' or 'Dwarf Queen,' two types that have proven to be wilt-proof.
- Moderate resistance was seen in Cultivars Bouquet Puff Mix 2080, Stokes Aster Standy Mix, Finest Mixed 684, and Astoria Mix Aster 2087.

Botrytis blight, *Botrytis cinerea*

Symptoms

- Fungal growth occurs on plant tissue that is dead, dying, or damaged.
- The petals start to get brown spots on them. It is possible for the entire flower to wither and die.
- Pests can also eat leaves and stems.
- To begin with, spots on young leaves are brown and found around the leaf 's periphery.
- These patches can spread and eventually cover the entire leaf, petiole, and stem.
- Fusarium spores, which look like fuzzy grey clouds, grow on decaying matter.
- Infections of stems can cause death of tissue above the infected area.
- If an infection spreads below the plant's root system, the entire plant could wilt and die.
- The root system of the plant is also vulnerable to fungus.

Favorable conditions

- Circumstances of cool wetness are favourable.
- The spores of the infection can best develop and disseminate in a humid setting.

Survival

- Remains of plants.

Management

- Don't water the lawn from above.
- Eliminate any decaying or wilting plant material (especially flowers) from the soil and surrounding plants.
- Having plants around helps with ventilation.

- Increasing greenhouse temperature and ventilation will help reduce humidity.
- Both Phyton 27 and Terraguard SC can be sprayed at rates of 1.3 to 2 fluid ounces per ten gallons of water.
- Prestop, a fungal bioagent made from the bacterium *Gliocladium catenulatum* strain J1446, should be sprayed at a concentration of 3.5 oz per 5 gallons of water.

Foot rot, *Phytophthora cryptogea*

Symptoms

- Identified by a rapid wilting of affected plants.
- This is a deterioration of the roots.
- The roots and lower parts of the stems are affected by the soft, damp rot.
- The browning occurs on the lowest part of the stems, the roots, and the leaves, making them look brownish black.
- Seedling damping off is also an effect.
- It's possible that the stem will wither and break off at or near the ground level.
- Soft, water-soaked rot could potentially harm the roots.
- Roots with disease may cause healthy roots to sprout nearby.
- The outer bark of the roots sloughs off and stays in the soil, making it easy to pluck diseased plants out of the ground.

Favorable conditions

- The conditions necessary for the disease's spread are present when there is an abundance of moisture, inadequate drainage in the soil, and chilly temperatures.
- Thriving in saturated, thick soils.

Survival

- Keep alive in dirt.

Management

- Keep away from planting in waterlogged soil. Only use soils that drain effectively.

- Build up planter beds.
- Don't water too much, and keep the nozzles of hoses off the floor.
- Mefenoxam should be sprayed on young plants.

Rust, *Coleosporium asterum*

Symptoms

- On the undersides of leaves, orange pustules filled with powdery spores begin to develop.
- The leaves of severely affected plants become yellow and eventually die.

Favorable conditions

- Invigorated by the free moisture provided by rain, dew, or fog.

Management

- Separate seedlings from the main crop as they develop.
- Don't water the lawn from above.
- Keep asters and cacti apart (it is from them the spores of rust fall on the asters).
- Start treatment as soon as notice rust and keep going until the weather stops being ideal for the disease. The most effective suppression was seen with Triforine (as Funginex), followed by Triadimefon (as Bayleton), and finally Mancozeb (as Fore).
- Spraying asters with a 1% solution of Bordeaux mixture is a good prophylactic treatment.

Root and stem rot, *Rhizoctonia solani*

Symptoms

- The disease initially manifests itself on the lowest leaves that are in direct touch with the ground.
- Infected leaves have a distinct water-soaked patch that spreads swiftly in cool, damp weather.
- In its early stages, the leaf is a deep emerald green; as it dries out, it eventually turns a rich chocolate brown.
- In some cases, the root or stem may be targeted as well.

- At first, the diseased region will look waterlogged; subsequently, it will sink and turn a dark reddish brown.

Favorable conditions

- Disease advances rapidly during cool and damp environments.

Survival

- This pathogen lives in dead plant matter and spreads through the soil.

Management

- It is important to have well-drained beds.
- Add PCNB or *Trichoderma* spp. to the top inch of soil before planting or transplanting.
- Thiophanate-methyl, iprodione, or *Trichoderma* spp. sprayed at the root of seedlings.

Aster yellows, *Phytoplasma*

Symptoms

- In most cases, new yellow leaves will appear during the vegetative growth stage.
- The clustering of leaves causes the leaf petiole to twist and stand upright.
- The disease causes the plant to cease developing and eventually die.
- Some flowers get unusually green petals as they reach a later stage of development.
- They look like malformed flowers.
- Once in a while, every flower will be a vibrant shade of green.

Survival and spread

- In warm climates, phytoplasma is spread by leafhoppers, propagative portions (bulbs, corms, and tubers), and overwintering weed and crop plants.
- The phytoplasma that causes aster yellows is spread by leaf hoppers.

Management

- Remove and destroy any diseased plants.
- Controlling the growth of grass and weeds in the flower bed to reduce leafhopper populations.

» Carbofuran (applied as a soil spray of Furadan 3G at 10 kg/ha), Imidacloprid (applied as a foliar spray of Confidor 200SL at 0.5 ml/L).

References

Sridhar, V., Naik, S. O., Swathi, P., & Mani, M. (2022). Pests and Their Management in Ornamental Plants: (Rose, Jasmine, Chrysanthemum, Crossandra, Marigold, Tuberose, Carnation, China aster, Gerbera, Gladiolus, Hibiscus, etc.). Trends in Horticultural Entomology, 1189-1237.

Beaumont, A. (1950). Diseases of China Asters. Gardeners' Chronicle, 128(3325).

5 Chrysanthemum

INSECT PESTS

Aphid, *Macrosiphoniella sanborni*

Damage symptoms

- Adults and juveniles alike feed on the cell sap of developing stems and the undersides of leaves.
- The plant loses its vitality and its leaves become yellow and fall off prematurely as a result of the damage.
- Aphid honeydew promotes the growth of sooty mould on affected leaves.
- Floral bases are also prone to aphid infestation.
- The size of flower buds diminishes and they never bloom on heavily affected plants.

Pest identification

- Nymphs have a dark, almost greenish appearance.
- The mature population is a rich chocolate brown.

Management

- Pongamia or Neem oil spray, 20 ml per litre.
- Applications of 4% Pongamia kernel extract or 1% Neem oil, followed by sprays of Dimethoate 30 EC @ 2 ml/L or Oxydemeton Methyl 25 EC @ 2 ml/L or Imidacloprid 200 SL @ 0.4 ml/L or Acephate 75 SP @ 1.5 g/L.
- Inoculation with 3 g/L of *Verticillium lecanii* (trade name "Microgermin").

- Concealment of a One-Day-Old Larvae of the *Chrysoperla carnea* at a ratio of 50:1 (aphid: chrysopid).

Thrips, *Microcephalothrips abdominalis; Frankiniella* spp.

Damage symptoms

- Thrips are typically found in or around flowering plants or new plant growth such as buds and young leaves.
- Thrips feed on soft, recent growth or blossoms by rasping or scraping the surface cells and sucking the cell contents due to the inability of their rasping and sucking mouth parts to penetrate deeply into plant tissue.
- Adults and nymphs alike eat on young, vulnerable leaves, causing them to become silvered, mottled, and distorted.
- Burning causes flowers to lose their colour, wilt, and dry up.
- It may take time for the damage to become apparent after feeding, but it will show up in the distorted growth of the damaged flowers, leaves, or fruit.

Pest identification

- Nymphs have a whitish appearance and a slim build.
- People of adult age are black in appearance.

Management

- In order to prevent thrips from damaging flowers, immature buds can be covered with perforated polythene bags before they bloom.
- Applying a spray solution of 0.05% dimethoate, 0.05% oxademeton methyl, 0.05% phenthoate, or 0.01% acephate.
- Pupal stages can be eradicated by raking the soil and applying Chlorpyriphos 20 EC @ 5 ml/L.
- Spraying 0.05% solutions of Cartap hydrochloride, Ethofenprox, or Fipronil may be necessary if the outbreak is particularly severe.
- Spinosad spraying for thrips control.
- Predatory little pirate insect (*Orius insidiosus*) nymphs and adults released.
- Planting disease-resistant cultivars such as Chandrika, Bangalore Local Yellow Double, M-7, Pankaj, and Yellow Star.

Leaf miner, *Liriomyza trifolii*

Damage symptoms

- The months of March through June are the worst for leaf miner infestations.
- The adults lay their eggs on the undersides of leaves, and the developing larvae mine the leaves for mesophyll, leaving behind telltale striations.
- Mined heavily, the leaves of a plant with severe damage will eventually become brown and dry.

Management

- Take off and throw away mined leaves that are severely damaged
- To catch the adult population, set up yellow sticky traps covered with Chlorpyriphos 20 EC @ 2.5 ml/L.
- To remove leaf miners, spray them with insect growth regulators such cyromazine, methoprene, fenoxycarb, and avermectin.
- Margosan (neem formulation) 3% dipped cuttings for 2 hours, followed by a 0.4% foliar spray to kill off any remaining pupae and adults.
- Spraying with a foliar systemic pesticide like Acephate or Spinosad may be necessary if the damage is extensive.
- To remove larvae, spray Triazophos 40 EC at 1.5 ml/L, Methomyl 40 SP at 1.5 g/L, or Abamectin 1.9 EC at 0.4 ml/L.
- For controlling adults, use a spray containing either Dichlorvos 76 EC at 1 ml/L or Decamethrin 2.8 EC at 1 ml/L.
- Parasitoid *Diglyphus* spp., released at a rate of 1,000 per week, can effectively control the larval stage.

Bud borer, *Helicoverpa armigera*

Damage symptoms

- Bud worm larvae are mostly herbivores, consuming only the terminal shoots, young foliage, flower buds, and flower heads of plants.
- The larvae make a tunnel to the growing floral organs and feast on them.
- Round holes, a *telltale sign* of feeding damage, can be found in flower buds and blossoms.

Pest identification

- Adults are around the size of human thumbs and are a medium brown colour with a distinct V-shaped speck on each of their wings.
- The male has green forewings, and the female has brown ones.

Management

- Cleanup on the battle field.
- Buds and blooms should be gathered and thrown away if they are damaged.
- Capturing and killing off the adult larval stage.
- Helilure pheromone traps should be set up at a rate of 15 per hectare.
- The crop can be protected from borer damage by spraying NSKE at a concentration of 4%.
- To apply 1% Neem oil spray.
- Promote the growth and reproduction of *Campoletis* and *Chelonis* parasitoids.
- Apply 500 LE, or 1.5 x 10^{12} POBs/ha, of HaNPV to the sprayer.
- To spray, use either 0.05% Methyl Parathion or 0.01% Ethofenprox.

Two spotted spider mite, *Tetranychus urticae*

Damage symptoms

- Mites are commonly observed on the upper surface of leaves, and in extreme cases of attack, the population extends throughout the entire plant.
- The worst of the damage occurs between the months of April and July.
- The active stage of the red spider mite is marked by a prodigious production of webs.
- As a result, the affected leaves turn brown, wilt, and eventually fall off.
- Failed new growth wither and die.

Management

- Infested shoots are cut off and burned.
- Dicofol 18.5 EC at 2.5 ml/L, Dimethoate 30 EC at 2 ml/L, Triazophos 40 EC at 1.5 ml/L, and Profenophos 50 EC at 1 ml/L are all effective as sprays.
- Spray with 1% Pongamia or Neem oil again after 15 days.

DISEASES

Rust, *Puccinia crysanthemi*

Symptoms

- In the early spring, rust can be very devastating.
- Up to 5 mm in diameter, yellow to tan patches appear on the upper leaf surface during infection, with the centres eventually turning brown.
- Disease symptoms manifest as elevated brown blister-like pustules on the undersides of leaves, which eventually turn white.
- Fallen leaves and fewer flowers.
- Brown, powdery spores are released in large quantities as these burst open.
- Infected plants lose strength and may not blossom at all.

Management

- The spread of a disease can be slowed by promptly eradicating any contaminated leaves or plants.
- Using a spray of Karathane at 0.025% or Wettable Sulfur at 0.3% on the plants helps keep the disease at bay.
- Fungicides containing Azoxystrobin 0.1%, Difenconazole 0.05%, Chlorothalonil 0.2%, or Thiophanate Methyl @ 0.1% are used to prevent plant disease.
- Powdery mildew can be avoided using chrysanthemum varieties like Achievement, Copper Bowl, Miss Atlanta, Powder Puff, and Orange Bowl.

Septoria leaf spot, *Sepotria chrysanthemella*

Symptoms

- During the cool, damp months of the rainy season, spots appear on leaves.
- As the rain splashes the germs around, they infect the lowest leaves first.
- Spots of varying sizes and shapes that are blackish brown with a yellow halo.
- They join forces to produce larger patches that eventually spread across the leaf's surface.
- Premature leaf withering caused by a severe infection causes the leaves to hang limp from the stem for a while before falling off.

» Flower buds get infected with the fungus just before they begin to bloom, and they rot away totally.

Favorable conditions

» It seemed like most of the infections originated from trash in the ground.

» The disease thrives in chilly, rainy climates, making it more dangerous during and after the monsoon.

» It does not flourish in warm climates.

Survival and spread

» Evidence suggested that infected soil debris was the predominant source of infection or that suckers were the primary vector for systemic infection.

» Splashing water disperses the fungus' spores.

Management

» It's important to clean up and dispose of any plant matter that may have been contaminated with a disease.

» In order to save water, it's best not to water too much.

» Beginning at the end of July, spraying regimens may include Benomyl (0.1%), Mancozeb (0.2%), Copper oxychloride (0.3%), Azoxystrobin (0.1%), Chlorothalonil (0.2%), Thiophanate methyl (0.1%), and Captafol (0.2%).

» Some chrysanthemum varieties are particularly resistant to disease and pests, including the Alpana, Aparjito, C.L. Philips, Flirt, Lilliput, Phillies, and Sarad varieties.

Alternaria leaf spot, *Alternaria* spp.

Symptoms

» Small dark brown patches on a yellow background are caused by a pathogen.

» Spots combine, and the decay of the leaf begins.

» When an infection is particularly severe, the leaves wither and eventually fall off the stem.

» The entire leaf eventually withers and dies.

Management

» The mixture of 2 g/lit Mancozeb and Copper oxychloride will be sprayed.

Powdery mildew, *Erysiphe cichoracearum* (*Oidium chrysanthemi*)

Symptoms

- In elder plants, infection is worse in damp situations.
- An ash-gray, powdery white growth of the fungus can be seen on the upper surface of the leaves.
- Leaves pucker or twist.
- Infected leaves wilt and eventually fall off the plant.
- Infected plants fail to grow and blossom.

Favorable conditions

- Climates that are both hot and humid, like those that promote the spread of this disease, are ideal.

Management

- Maintaining low relative humidity, adequate ventilation, and adequate spacing are all suggested.
- Spraying at 10- to 15-day intervals with 0.2 percent Wettable Sulfur, 0.03% Triforine, 0.05% Thiophanate-methyl, 0.05% Dinocap, 0.1 percent Cabendazim, 0.01% Benomyl, or 0.2 percent Captan is an effective method of disease prevention.

Fusarium wilt -*Fusarium oxysporum* f. sp. *chrysanthemi*

Symptoms

- Early symptoms include stunting, yellowing of leaves, and wilting, most typically on just one side of the plant.
- The affected leaves gradually fall off the plant starting at the roots. Virus infections cause stunting and, in many cases, the complete absence of flowering in the affected plants.
- Vascular discolouration in the stems appears reddish brown.
- Leaves on plants may darken and die if they are under water stress.
- If the plant wilts, the root or stem base could rot.

Favorable conditions

- Warmth, high relative humidity, excess water, and inadequate drainage all promote the growth of pathogens.

Survival and spread

- In this case, the fungus may be traced back to the ground.
- As a result of infected cuttings and contaminated soil, the disease rapidly spreads.

Management

- Cuttings are the most common way for the disease to spread, thus it is crucial that you use healthy cuttings when planting.
- Neutralize acidity by applying nitrate nitrogen fertiliser and bringing the pH up to 7.0.
- Rooted cuttings are dipped into a Thiram solution (1.5 g/L water) before being planted.
- Copper oxychloride at a concentration of 2.5 g/L, Trifloxystrobin and tebuconazole at 0.75 g/L, difenoconazole at 0.5 ml/L, or carbendazim at 0.1 percent are all effective drench treatments for infected soil.
- Strict hygiene, regular monitoring, rotating crops, and removing any sick plants from the field (known as "rogueing") will further reduce the spread of disease.

References

Khan, A. U., Choudhury, M. A. R., Khan, A. U., Khanal, S., & Maukeeb, A. R. M. (2020). Chrysanthemum production in bangladesh: significance the insect pests and diseases management: A review. *Journal of Multidisciplinary Applied Natural Science.*

Visser, P. B., Maagd, R., & Jongsma, M. A. (2007). III3 Chrysanthemum. Biotechnology in Agriculture and Forestry, 61, 253-272.

Shimira, F., Uğur, S., Özdemir, Ş. M., & Mendi, Y. Y. (2021). Future and Prospect use of Pyrethrum (Chrysanthemum cinerariifolium) as Part of the Integrated Pest and Disease Management (IPDM) Tool in Turkey. Turkish Journal of Agriculture-Food Science and Technology, 9(1), 150-158.

Daughtrey, M., & Buitenhuis, R. (2020). Integrated pest and disease management in greenhouse ornamentals. Integrated pest and disease management in greenhouse crops, 625-679.

6 Crossandra

INSECT PESTS

Brown scale, *Saissetia nigra* and White scale, *Orthezia insignis*

Damage symptoms

- Adult scales, which are brown in colour, can be found on the petiole and stem, as well as under the leaves.
- Due to the severe infestation, the plant's growth is stunted, and its leaves turn yellow and fall off.
- White scales feed on the sap of young, vulnerable branches.
- Infested plant parts turn yellow and wilt as they lose their vitality.
- *Orthezia* sp. honey dew encourages the establishment of sooty mould, which stunts plant development.

Management

- Any infected parts of plants should be cut off and burned to stop the spread.
- Carbofuran 3 G was applied at a rate of 1.0 kilogramme active ingredient per hectare.
- It is recommended to use a mixture of Malathion 50 EC at 1 ml/L and either Ethion 50 EC at 1 ml/L, Chlorpyriphos 20 EC at 2.5 ml/L, or Dimethoate 30 EC at 2 ml/L.
- Apply 1% Pongamia oil after spraying the aforementioned chemical.
- Combination of 1% Pongamia oil and 1% Neem oil with 0.5 % Methomyl and 0.05 % acephate.

Spike borer, *Helicoverpa armigera*

Damage symptoms

- The young larvae eat holes in the spikelets.
- Damage is done by the mature larvae, which feed on the developing spikes, flower buds, and seeds.
- Spikelets are frequently webbed with a thin, silky web.
- More than half of the spikelets in a damaged spike have visible boreholes, and the overall damage to the spike can be as high as 20%.

Pest identification

- Greenish caterpillars measure about 2.5 cm in length; they have a black head and a pale, yellowish white early. Grayish brown or greenish cylinders with yellow lines along the sides and a few hairs here and there can be found laterally.
- Small and a pale brown, the pupa is a nocturnal insect.
- Butterflies with brownish-colored wings. The wings of an adult moth measure between 30 and 45 mm in length. The forewings of this insect are either brownish or reddish brown (females) or dull green to yellow or light brown, while the wings of the insect's hind legs are white with a broad black outer edge.

Management

- Borer populations can be controlled by collecting and killing mature larvae.
- Set up 10-15 pheromone traps per hectare.
- At 15-day intervals, spray either 0.01% fenvalerate or 0.05% chlorpyriphos, then 0.05% methyl parathion or 0.07% endosulfan.
- To prevent spikes from being eaten by borer, spray them with a mixture of Endosulfan 35 EC @ 2 ml/L or Methyl parathion 50 EC @ 1 ml/L and Diflubenzuron 25 WP @ 2 g/L two to three times at 15-day intervals.
- Inject HaNPV.

Midge, *Contarinia* sp.

Damage symptoms

- Galls are abnormal growths on plants caused by larvae feeding on the plant's tissue and releasing toxins.

- Maggots are parasites that feed on the ovaries of growing buds and then pupate inside of them.
- As a result of the damage, the buds will not open and will eventually decay.

Pest identification

- Maggots have a distinct orange hue.
- The adults of most fly species are black. There are many species of fly, and all of them are tiny, with lengths of less than 1 mm. Their large antennae and hairy wings distinguish them.

Management

- One kg of active ingredient per hectare of Carbofuran 3 G applied to the soil.
- When a few buds start to rot, spray them with Triazophos 40 EC at 1.5 ml/L, Ethion 50 EC at 1 ml/L, Acephate 75 SP at 1.5 ml/L, or Chlorpyriphos 0.05%.

Tussock caterpillar, *Dasychira mendosa*

Damage symptoms

- Defoliation is caused by larvae feeding on leaves.

Pest identification

- The larva has black prothoracic and pre-anal tufts and greyish brown body coloration; the prolegs are a bright red.
- The cocoon's developing pupa.
- The full-grown moths are a drab brown with yellow highlights.

Management

- To rid the plant of the larvae and any affected sections, collect them.
- Apply a spray of 1.5 ml/L of Malathion or Endosulfan 35 EC.

Aphid, *Aphis craccivora*

Damage symptoms

- Nymphs and adults both feed on the sap of young leaves and blossoms.
- The foliage loses its moisture and dries up.
- Botanical buds with pests inside them don't bloom.

Pest identification

- A nymph who is powdered and a drab grey.
- Only seldom can adult females grow to be longer than 2 mm, and their lustrous black apterous dorsal bodies characterise them as adults.

Management

- One and a half ml per litre of Dimethoate should be sprayed.

DISEASES

Fusarium wilt, *Fusarium oxysporum*

Symptoms

- Wilt appears to be localised.
- A month after being transplanted, the disease has already appeared in the wild.
- The leaves of diseased plants turn yellow and droop.
- Discoloration around the leaf margins appears pinkish brown.
- In a week to ten days, the discolouration will have reached the midrib.
- The top of the stem dries up.
- This condition is characterised by dark lesions on the roots that progress to the collar area, causing the cortical tissue to flake off.

Survival and spread

- The chlamydia spore can live in the soil for a long time.
- They are carried by the water used for irrigation.

Management

- In order to reduce the spread of the disease, infected plants must be removed and killed.
- Soil should be drenched with Captan (0.2% solution) before planting.
- Carbendazim 0.1% or copper oxychloride 0.25 % drenched into the soil once every 30 days is sufficient to prevent the disease from spreading.
- If necessary, the treatment can be repeated after 3 to 4 weeks.
- *Trichoderma viride* strain Tv1 conidial suspension applied to soil at a concentration of 1×10^{11} cfu/ml.

Leaf blight, *Alternaria alternata*

Symptoms

- Yellow dots, often uneven in shape, appear on the upper surface of infected leaves.
- As time passes, the necrotic tissue around them swells, turns brown, and forms concentric dark brown rings.
- Yellowing and early fallout of infected leaves are classic symptoms of this disease.

Management

- Pay attention to hygiene in the field.
- In the spring and summer, spray plants with Mancozeb (0.2%) or Bavistin (0.1%) every 15 days.
- Benomyl 0.1% spraying (or) Use either 0.2% Mancozeb or 0.1% Carbendazim.
- At 30 and 60 days after planting, spray 2 g of *Bacillus subtilis* per litre of water.

Root rot, *Macrophomina phaseolina*

Symptoms

- The most noticeable symptoms are leaf yellowing, root discoloration and rot, and dark brown lesions on the stem. Tuber decay is another possibility.
- Sclerotial bodies, which appear as tiny black dots, can be detected in the stem tissue.

Management

- *Pseudomonas fluorescens* @ 2 g/L of water was used to treat the tubers for 20 minutes.
- 30 days after planting, saturate the soil with 2 g of *Bacillus subtilis* per litre of water.

References

Jegathambigai, V., Wijeratnam, R. S. W., & Wijesundera, R. L. C. (2009). Control of Fusarium oxysporum wilts disease of Crossandra infundibuliformis var. Danica by Trichoderma viride and Trichoderma harzianum. Asian Journal of Plant Pathology, 3(3), 50-60.

Sridhar, V., Naik, S. O., Swathi, P., & Mani, M. (2022). Pests and Their Management in Ornamental Plants: (Rose, Jasmine, Chrysanthemum, Crossandra, Marigold, Tuberose, Carnation, China aster, Gerbera, Gladiolus, Hibiscus, etc.). Trends in Horticultural Entomology, 1189-1237.

Reddy, P. P., & Reddy, P. P. (2014). Ornamental crops. Biointensive Integrated Pest Management in Horticultural Ecosystems, 185-200.

7 Gerbera

INSECT PESTS

Leaf miner, *Liriomyza trifolii*

Damage symptoms

- At the three- to four-leaf stage, the problem first appears in newly planted crops.
- The larvae cause damage to the palisade mesophyll inside the leaves, which reduces photosynthesis and yield, and thus the quality of the produce that may be sold.
- In the event of a strong attack on the plants' growth, the leaves will become brittle and dry and will be covered in large mining scars.
- Those plants develop growth defects and their blossoms are small and fragile.
- We need to cut back on the flowers.

Pest identification

- The larvae have a bright orange-yellow body.
- The adult version of this species has a brown body, black and yellow stripes, and a yellow head. The length of these items is between 2.0 and 2.5 mm.

Management

- Mined leaves should be thrown away or burned.
- To provide heavy defence and promote parasitization by *Diglyphus isaea*, set up yellow sticky traps coated with Chlorpyriphos 20 EC at a concentration of 2.5 ml/L.

» Plant a trap crop such as field bean (*Vicia faba*) as an intercrop.

» Use a spray with 1% Pongamia oil or 4% Neem seed kernel extract.

» Apply a spray of Triazophos 40 EC at 1.5 ml/L or Abamectin 1.9 EC at 0.4 ml/L to remove larvae.

» For controlling adults, use a spray containing either Dichlorvos 76 EC at 1 ml/L or Decamethrin 2.8 EC at 1 ml/L.

» Diffusing 1000 of the parasitoid eulophid *Diglyphus intermedius* every week.

Whitefly, *Bemisia tabaci*

Damage symptoms

» Discoloration, wilting, and drying of leaves are the results of nymphs and adults sucking cell sap from the underside of leaves.

» In a severe infection, the nymphs' excreted honeydew will become black with sooty mould.

Favorable conditions

» The pest is most active when temperatures and humidity.

Management

» Avoid places that are too dark or too wet.

» Those leaves with the most pest infestations should be thrown in the fire and burned.

» Put up sticky yellow traps to catch adults.

» Alternate spraying Pongamia oil 1% with Triazophos 40 EC @ 1.5 ml/L, Methomyl 40 SP @ 1.5 g/L, or Acephate 75 SP @ 1.5 g/L for nymph control.

» Adults can be sprayed with a solution of Dichlorvos 76 EC at a concentration of 1 ml/L.

» Let the *Encarsia formosa* parasitoid larvae loose in the wild.

Thrips, *Thrips palmi*

Damage symptoms

» From July to February, thrips are a problem because they feed on the sap of plants.

- Curling, brittle, and dark spots appear on damaged leaves.
- Buds and flowers affected by the disease are discoloured and deformed in shape, reducing both their aesthetic value and their potential yield.

Management

- Spray Pongamia oil 0.5% twice or thrice every 15 days, when a new flush begins, along with Acephate 75 SP at 1.5 g/L or Dimethoate 30 EC at 2 ml/L.
- In cases of severe infestations, Fipronil 5 SC @ 1.5 ml/L should be used.
- Chlorpyrifos 20 EC @ 5 ml/L should be sprayed heavily on the ground to kill pupae.

Aphid, *Myzus persicae*

Damage symptoms

- In the fall and winter, most of the attacks occur on the blossoms and buds.
- Floral damage, including browning, withering, distortion, and wilting, is caused by nymphs and adults colonising on the underside of petals, then spreading to the rest of the flower and sucking the sap.
- Aphids' honeydew turns sticky and black when sooty mould develops on it, making the affected plant ugly and its blooms unsellable (photosynthetic activity of plant is hindered).

Management

- In the early stages of an infestation, sprinkle Pongamia or Neem oil 1% sprays.
- In a spray, apply either Dimethoate 30 EC at 2 ml/L or Oxydemeton methyl 25 EC at 2 ml/L every 10 to 15 days.
- Spray Imidacloprid 200 SL @ 0.4 ml/L, Cartap hydrochloride 50 SP @ 1.5 g/L, or Acephate 75 SP @ 1.0 g/L if the infestation is really bad.
- The aphid population can also be decreased by using *Verticillium lecanii* at a concentration of 3 g/L.

Broad mite, *Polyphagotarsonemus latus*

Damage symptoms

- Brown dots form on the leaves where the mites have been sucking the sap.

- The growth of infected plants is stunted.
- Leaves are bending downwards.
- Leaves and flower buds are damaged in a greenhouse.
- Flowers suffer irreparable damage and lose their marketability when subjected to a severe attack.
- When there are a lot of mites, a web-like structure develops.

Favorable conditions

- Conditions of high heat and low humidity are particularly conducive to its spread.

Management

- Gather infected leaves and dispose of them.
- Apply an oil spray with 1% Annona, Pongamia, or Jatropa.
- Use a mixture of 0.15% dichlorvos, 0.1% methomyl, 0.0025% vertimec, 0.025% diafenthiuron, 0.05% amitraz, and 0.015% flufenuxuron in a spray bottle.
- Dicofol (1.5 ml/L) or Endosulfan (2.0 ml/L) can be sprayed.
- Apply 0.4 ml/L of Abamectin 1.9 EC or 1 ml/L of Propargite to a spray bottle filled with water.

DISEASES

Root rot, *Pythium irregulare, Rhizoctonia solani*

Symptoms

- The disease caused a slowing of development.
- A brown lesion can be seen on the lower stems.
- The root skin can be readily peeled off.
- Root rot eventually kills the entire plant.
- It's death time for the plants.
- It's important to note that older plants are more vulnerable to assault by *Rhizoctonia solani,* which leads to greater losses.

Management

- Grow plants in sterilised soil.
- Do not over-water.
- A fungicide, such as Copper oxychloride at a concentration of 2 g/L in the soil, may be useful in preventing the spread of the disease.
- Soak in a 2 g/L solution of Benlate, Bavistin, or Topsin/Roko.
- *Pseudomonas fluorescens* water at a concentration of 2 g/L.

Crown rot, *Phytophthora cryptogea*

Symptoms

- Infection starts at the collar of the stem, right where it meets the soil.
- Light brown, uneven, water-soaked patches on flower stalks and petals are the telltale sign of this disease.
- The spots multiply quickly and join together to produce larger, flatter lesions.
- Sometimes the foliage will have a yellowish hue.
- Plants suddenly wilt as their roots decay.
- The plants' crowns become dark and decay.
- The foliage and flora wither and die.
- The infections worsen in damp environments, spreading throughout the entire flower head and ultimately causing blossom blight and stalk rot.

Favorable conditions

- Drizzle and chilly, damp weather is ideal for the spread of the disease.

Survival and spread

- The fungus is transmitted by soil and infects plants when their roots come into contact with the ground.

Management

- The fungus is transmitted by soil and infects plants when their roots come into contact with the ground.
- Solarization successfully eliminated the pathogen in soil to a depth of 30 centimetres.

» Crop failures are less common when disease-free soil is used.

» The affected blossoms need to be gathered and thrown away.

» It's best not to water plants too much.

» At a dose of 100 ml/m2, Vapam soil fumigation is highly effective.

» Reducing the disease's prevalence is soil warming (26° C).

» Metalaxyl soil drenching or 0.2 g/L contact fungicides like Captan or Copper oxychloride.

Blight / Grey mold, *Botrytis cinerea*

Symptoms

» Long brown dots extend from the petioles.

» The foliage dies and turns yellow.

» Fungus eats its way through flowers, leaving grey patches on the petals and rotting the hearts.

» The ground-level stems are annihilated.

» Gray fungal growth spreads across infected tissues.

» Young, developing tissues are particularly vulnerable, and the fungus can quickly destroy them.

» Small black dots appear on the ray florets of Gerbera flower heads that have been grown in humid conditions.

Favorable conditions

» Typically takes place early in the morning when the relative humidity is above 90% for two hours.

» The ideal conditions for germination and growth of the spores are damp leaves and a temperature between 15 and 25 °C.

Survival and spread

» Resting structures include mycelium and sclerotia, which are found in decomposing plant matter or on wild plants (Amaranthus Genus).

» The pathogen can be dispersed through the air and water.

Management

» Remove weeds and space plants in the polyhouse so that air can flow freely.

- Keep the relative humidity to a minimum.
- The late afternoon is not a good time to water.
- Clean up the crop area. Reducing the spread of disease by removing and destroying contaminated tissue.
- Plants are more likely to become infected if they are planted deeply, have poor drainage, and/or lack enough ventilation.
- Apply a spray of either 1.5 g/L Dithane M-45 or 0.5 g/L Quintol.

Powdery mildew, *Erysiphe chichoracearum, Oidium erysiphoides* f. sp. *gerbera*

Symptoms

- Infected plants will develop a white powdery coating on their leaves and blossoms.
- When under attack, the leaves will curl.
- A brown, scab-like lesion can appear on some gerbera plants.
- Gerbera plants that contract powdery mildew swiftly wither and die.

Favorable conditions

- When the conditions of high humidity, moderate temperatures, and cloudy skies persist for an extended length of time, this fungal disease thrives.

Spread

- The microorganisms that cause powdery mildew in gerberas can travel via the air to infect other plants.

Management

- Any infected plants or leaves should be thrown away immediately.
- Drip irrigation systems provide gentle watering, protecting plants from humid conditions.
- Use a spray solution of 2 g/L of Wettable Sulfur, 1.5 g/L of Diathane M-45, or 0.5 g/L of Quintol.
- Spraying 0.5 ml/L of Fenarimole or 0.3 ml/L of Dianocab.

Stem rot, *Fusarium solani*

Symptoms

- As the plant dies, the petiole of the leaves turns black at their bases.
- Black vessels are seen in the leaf stem when it is cut.

Management

- Start seeds in sterilised potting soil.
- Do not keep diseased plants.
- Use 2 gm per litre of Topsin/Roko, 2 gm per litre of Aliette, 1.5 gm per litre of Kavach, or 1.5 gm per litre of Kocide as a spray.

Leaf spot, *Alternaria* sp.

Symptoms

- Leaf symptoms appear when water sits on the leaf surface for an extended period of time.
- On the florets, brown spots had started to appear.
- The leafs become covered in black, round dots.
- Leaf dots develop white centres.

Management

- Keep the relative humidity low and avoid watering the leaves directly.
- Apply a mixture of 1.5 g/L Diathane M-45 and 0.5 g/L Quintol by spray.

References

Reddy, P. P., & Reddy, P. P. (2016). Gerbera. Sustainable Crop Protection under Protected Cultivation, 355-362.

Chudali, B., Shrestha, S., Rajbhandari, B. P., & Sah, L. P. (2020). Status of insect pests and diseases associated with gerbera in Kathmandu valley. Nepalese Journal of Agricultural Sciences 2020, volume 19, 82.

Deng, Z., & Bhattarai, K. (2018). Gerbera. Ornamental crops, 407-438.

Moyer, C., & Peres, N. A. (2008, December). Evaluation of biofungicides for control of powdery mildew of gerbera daisy. In Proceedings of the Florida State Horticultural Society (Vol. 121, pp. 389-394).

8 Gladiolus

INSECT PESTS

Thrips, *Taeniothrips simplex*

Damage symptoms

- Damage to leaves and spikes is caused by both the yellow nymphs and the adults, which are black and seem like they've been dipped in acid.
- When the attack is severe, the leaves and spikes become silvery-streaked, brown, distorted, and dry.
- The gladiolus corms been storing will be ruined, and late-blooming plants in general will suffer.
- Stunted flower spikes, unopened or deformed blossoms, and tiny white spots all over the petals are all signs of a fungus infection.
- The corms of infected plants get sticky, dry out, and generate sickly sprouts.

Pest identification

- Eggs laid in the leaf tissue and corms are opaque white, smooth, and bean-shaped.
- Nixie hues of yellow and orange.
- White and milky as adults, then turning brown. A faint transverse band can be seen on the underside of the wings.

Management

- The acorms are dipped in a mixture of Primophos methyl (0.1%), Acephate (0.15%), and Cypermethrin (0.05%).

- Two to three applications of Acephate 75 SP at a rate of 1.5 g/L, Dimethoate 30 EC at a rate of 2 ml/L, or Oxydemeton methyl 25 EC at a rate of 2 ml/L should be made at 10-day intervals.
- Corms infested with thrips can be entirely eradicated by storing them at 2 °C for six weeks and then treating them with hot water at 46 °C.

Cut worm, *Agrotis segetum*

Damage symptoms

- Cut worms are commonly seen in the crop during the first month.
- The female moth will deposit her eggs on lower plant sections.
- At night, larvae munch on tender new growth.
- Mature larvae with a claylike colour feed on plants by severing their roots.
- Up until the three-leaf stage, plants are susceptible to harm.
- Worms that burrow into the ground and eat the corms and developing spikes do a lot of damage.

Management

- Pupae are at risk when ploughing is done in the summer because they are exposed to predators.
- Rice bran, jaggery, and Chlorpyriphos make up the poison bait (10: 1: 0.5)
- The use of Methyl parathion 50 EC at 1 ml/L, Quinalphos 25 EC at 2 ml/L, or Accphate 75 SP at 1 g/L as a spray is recommended.

Leaf caterpillar, *Spodoptera litura*

Damage symptoms

- A clutch of ash-colored eggs is placed on the underside of a leaf.
- The first instar larvae scrape the underside of leaves for food.
- The most noticeable sign is leaf skeletonization.
- Mature greenish-brown larvae consume a lot of leaf tissue overnight.

Management

- Plowing the ground deeply in the summer can expose pupae to predators.
- Trap and destroy egg masses and leaves infested with first-stage larvae.

- Sprays containing Diflubenzuron 25 WP at 2 g/L and either Quinalphos 25 EC at 2 ml/L or Chlorpyriphos 20 EC at 2.5 ml/L.
- Insect damage caused by first-instar larvae can be reduced by spraying 1% Neem oil or 4% Neem kernel extract.

Mealy bug, *Ferrisia virgata*

Damage symptoms

- Underground corms in the field become infested first and then the problem spreads to the storage area.
- Corms are damaged by nymphs and adults because they drink the corm's sap, causing the corm to shrivel and die.

Management

- Every two weeks, spray either Methyl parathion 50 EC at 1 ml/L, Dimethoate 30 EC at 2 ml/L, or Acephate 75 SP at 1.5 ml/L.
- Apply a 2% spray of pongamia or neem oil.
- Mist the area with 3 g/L of *Verticillium lecanii.*

Aphids, *Aphis gossypi, A. craccivora*

Damage symptoms

- Aphids feed on the sap of afflected plants, weakening and disfiguring them.
- Honeydew is secreted, luring black mould and ants.
- Aphids are a prevalent pest throughout the months of February and March.

Management

- Once every two weeks, spray it with 0.1% Malathion or Rogor.

DISEASES

Fusarium wilt, *Fusarium oxysporum* f. sp. *gladioli*

Symptoms

- Newer leaves appear stunted, and older leaves begin to yellow and fall off.
- Also, the spike may not develop normally and may be a duller shade of its original hue.
- Blackened spots at the plant's base could even spread to the corm.

- The entire plant will begin to droop in due time.
- When the corm is cut open, the interior will have a marbled brown appearance.

Favorable conditions

- Temperature, nitrogen, anaerobic conditions, and carbon dioxide buildup are all conducive to fungal growth (CO_2).

Survival and spread

- Soil-borne and transmissible via corms, it may wreak havoc on plants year after year.

Management

- Infected corms should not be planted.
- It is recommended to soak corms or cormels in hot water (50 °C) for 30 minutes prior to planting.
- The corms should be soaked in a Benlate/Bavistin 0.02% solution for 30 minutes before being planted.
- Keep the soil pH between 6.6 and 7.0.
- For fertilisation purposes, nitrate is the best nitrogen source.
- At the time of planting, apply 3 gm of Furadan and 3 gm of Thimet per square metre of soil.
- Corms inoculated with specific isolates of *Fusarium moniliforme* (*Gibberella fujikuroi*) var. *subglutinans* and *Fusarium solani* were resistant to infection by *Fusarium oxysporum* f. sp. *gladioli* in field experiments.
- Effective wilt reduction (between 58 and 92%) was achieved when *T. harzianum*/*T. viride* was used alone or in conjunction with compost (mushroom, vermin, paddy straw).
- Applying 1.25 tonnes of Neem cake per hectare together with *Trichoderma harzianum* before planting and then spraying the soil with 0.1% Carbendazim one month after planting is an effective method of disease management under field settings.
- *T. virens* and Carboxin were successfully integrated into the corm treatment process to drastically minimise gladiolus corm rot or wilt.

Botrytis blight, *Botrytis gladiolorum*

Symptoms

- » There are huge, brown dots on the leaves, which range in shape from circular to round.
- » Spots can also be found on the stems and blooms.
- » The patches start out light brown but quickly darken.
- » After a downpour, florets may get a soft, brown rot at the base.
- » When left in a humid and cool environment, the fungus spores spread and cover the entire blossom in a brownish-grey growth.
- » Corms are another host for the infection.

Spread

- » The spores of the fungus can spread via wind and rain.

Management

- » Infected corms should not be planted.
- » It's important to remove and dispose of any flower spikes that have seen better days.
- » Effective pathogen eradication from planting stock can be achieved through the use of hot water treatment (52°C) in combination with Carbendazim (1 g/L water).
- » Carbendazim (1 g/L of water) and Mancozeb (2 g/L of water) should be sprayed on plants at all times to keep diseases at bay.
- » Dithane M-45 is dusted on healthy corms before they are stored.

Curvularia leaf spot, *Curvularia trifolli* f. sp. *gladioli*

Symptoms

- » Leaf sheaths, petal undersides, and flower petals all develop water-soaked oval to elongated brownish patches with a brown edge and a yellow halo.
- » In time, the white spots will become brown, and then black.
- » Within the expanding spots, concentric rings form.
- » A rotting corm is useless.

Spread

- The blight grew rapidly in wet, warm circumstances.

Management

- Infected corms should not be planted.
- Make sure to use pasteurised soil when planting.
- Fungicide use is recommended for plant safety. Mancozeb 0.2% sprayed every 10 days efficiently prevents the problem.

Rust, *Uromyces transversalis*

Symptoms

- When rust spores break through the surface of a leaf, they first show as tiny yellow spots, and then as pustules of orange rust.
- These initial orange pustules are subsequently encircled by black pustules that contain teliospores or resting spores later in the year.
- Typically stretched throughout the width of the leaves, the spore eruption moves transversely across the width of the leaves, and each spore can be up to 3 mm in length.
- When plants are severely injured, they may not blossom at all and/or produce mature corms.
- There has been an outbreak of a disease that can wipe out entire commercial gladiolus fields in nurseries.
- Rust can be managed with fungicides like Myclobutanil, Penconazole, or Triticonazole.

Favorable conditions

- If there isn't enough air flow, the humidity will help the fungus grow.
- Infection of gladiolus plants is at its peak when temperatures are between 10 and 20 °C and leaves are wet for at least 12 hours.

Survival and spread

- The pathogen can survive the winter in decaying plant matter.

Management

- To prevent spores from overwintering, rake up and burn all fallen leaves in the fall.

- As soon as notice diseased leaves, pick them off by hand.
- Prevent rust from infecting plants, water them directly instead of soaking the leaves.
- The plants will flourish them enough of air to breathe.

Scab, *Pseudomonas marginata*

Symptoms

- Yellowing occurs at the leaf tips, and areas of reddish brown coloration appear at the base of the plant.
- They are then transformed into vertical stripes on the leaves.
- The plant's leaves wilt and its stem cracks near the ground.
- The sick corms develop small, yellow, and then brown lesions with distinct borders.
- The lesions range in size from microscopic to a quarter of an inch in diameter; they are round in shape and water-soaked.
- These patches, ranging in hue from yellow to dark brown, exude a sticky gum.
- Lesions are relatively superficial.
- Plants with the worst symptoms may die, while those with milder symptoms still look unhealthy.
- Soil mites appear to be the carrier organisms in this case.

Management

- Infected corms should not be planted.
- Knives should be sterilised regularly.
- Continue efforts to keep insects and mites at bay.
- Use a watering method that prevents water from collecting on the leaves.
- Don't touch plants if they're moist.
- Where gladioli are grown, other, non-host crops should be rotated in every few years.

References

McCulloch, L. (1941). Gladiolus diseases and insects (No. 1860). US Government Printing Office.

Sowmya, D. S., & Rao, M. S. (2012). Combined effect of Pseudomonas putida and Paecilomyces lilacinus in the management of disease complex in Gladiolus grandiflorus L. Pest Management In Horticultural Ecosystems, 18(2), 204-209.

Reddy, P. P., & Reddy, P. P. (2016). Gladiolus. Sustainable Crop Protection under Protected Cultivation, 375-380.

Wang, C. L. (1982). Insect pests and their injuries on gladiolus in Taiwan. Journal of Agricultural Research of China, 31(2), 173-176.

9 Lily

INSECT PESTS

Aphids, *Neomyzus circumflexus*

Damage symptoms

» The crescent-marked lily aphid is a frequent pest of lily plants both in the wild and in greenhouses.

» Both nymphs and adults congregate on young, vulnerable parts of plants to feed on the sap.

» Flowers lose their vivid hues, buds drop early, and young shoots wither.

» They weaken plants physically by twisting leaves and distorting flower buds.

» Aphids of various species are responsible for the colonisation of lilies and the subsequent spread of viral infections.

Management

» Aphids can be managed to a limited extent with the help of insecticidal soaps. For best results, applications should be made on a consistent schedule.

» To combat aphids, growers can use a foliar spray or a soil drench containing a Neonicotinoid with the appropriate labelling.

» Successful aphid control using the insecticide acephate (e.g., Orthene) has been reported. When malathion is sprinkled on the soil and then watered in, it provides long-lasting protection against aphids.

» When administered at a dilution of about 1%, light summer oil is quite effective. Because of their low toxicity, oils can be used alongside pesticides and fungicides without worry. The best time to spray oil is soon before sunset.

- Aphids can be kept out of the greenhouse by installing screening with 300 holes per inch over the vents and doors.
- Any plant debris that could potentially house overwintering eggs must be removed and disposed of.

Beetle, *Lilioceris lilii*

Damage symptoms

- Grubs in their juvenile stages eat the undersides of leaves, leaving behind bare, white or brown areas.
- The elder grubs not only consume the petals, stem, and seed pods, but also consume entire leaves, beginning at the leaf tips and working their way back to the plant's centre.
- The adult beetles damage plants by boring round holes in the leaves, but they also eat flower petals and seed pods.

Survival and spread

- Adult red lily beetles hibernate in soil, leaf litter, or other protected locations.

Management

- All stages of a beetle's life cycle—from adults to larvae and eggs—can be manually eliminated.
- Apply an insecticide containing Pyrethrum (0.5%) or another approved pesticide to the adult insects. Rotenone shows promise as an alternative.
- Insecticides like Thiacloprid, Acetamiprid, Deltamethrin, or Lambda-cyhalothrin may be needed to control severe infestations.
- Pyrethrum, a natural pesticide, can be used by organic gardeners to remove the beetles' larvae but has less of an impact on the adults.
- Put to use an insecticide containing neem, such as Azatin, Aza-Direct, or BioNeem. Neem is most effective on very young (first instar) larvae.
- To eliminate the mature larvae that overwinter just below the earth's surface, a pesticide such as acephate (e.g., Orthene) or a neonecotinoid should be sprayed heavily over the soil.

Thrips, *Liothrips vaneeckii*

Damage symptoms

- It appears that thrips feeding is concentrated at the scale bases, where the bulb is severely weakened and becomes flaccid.
- Corm and flower malformations and discolouration in lilies (bulbs).
- Flower buds and blooms that have been attacked often fail to open or only open partially.
- Surfaces that have been infested have an unusual lustre and shine in a pale grey hue.
- Because of this, germs and fungi can easily enter and cause the bulb to decay.

Management

- To kill off the bug, soak the bulbs in water heated to 44 °C for an hour.
- Corms that have been infested should be stored at 2 °C for a period of six weeks.
- Benzene hexachloride should be used for dusting light bulbs (BHC).
- Insecticide solutions can be used to wash or dip bulbs.
- Apply 1 ml/L of methyl parathion or 2 ml/L of endosulfan as a spray.
- Acephate 0.1% spray should be applied every 10 days, on average.
- Before the buds begin to form, spray the plants with Malathion (1 millilitre) or Carbaryl 50 WP (2 gm per litre of water).

Weevil, *Agasphaerops nigra*

Damage symptoms

- The larvae eat their way through the lily 's stem and bulb.
- In March and April, adult weevils appear and begin their lives as leaf eaters.

Management

- Weevils can be effectively managed with systemic pesticides.

Bulb mites, *Rhizoglyphus echinopus*

Damage symptoms

- Bulb mites tend to thrive in warm regions, and their populations tend to be quite high, especially above the basal plate and in the spaces between the scales.

- They invade the bulb through the middle after assaulting the roots and the base.
- Bulbs and roots will be eaten by the pests.
- Burrow into the stems and infest the scales.
- The leaves and stems of infested plants turn a sickly yellow colour.

Management

- Prepare bulbs by soaking them in water heated to 44 °C for one hour.
- Put the dry bulbs in an airtight jar with some Para dichlorobenzene (the active ingredient in moth balls) for 12 hours.
- Sprinkle some sulphur on flower bulbs to get them started.
- Burn any light bulbs that are severely contaminated.
- Lamps can be sterilised in steam at 60 °C for 30 minutes, or at 80 °C for 20 minutes.
- Rice can be grown in rotation with other crops, allowing soils to dry out from the sun, and sesbania can be ploughed under water for extended periods of time.
- Dicofol at 0.025-0.04% and Monocrotophos at 0.04% should be sprayed every other week.

Slugs and snails

Damage symptoms

- When lily shoots first appear, slugs and snails feast voraciously on them.
- At the end of the season, they can climb the stems and totally strip them of leaves.

Management

- Weeding out unwanted vegetation and replacing it with mulch or stone will help keep the area around lily beds more manageable.
- Regular cultivation of the top few inches of soil expose the eggs to the elements (including frost and birds), ensuring their demise.
- In the late afternoon, scatter bait containing Metaldehyde among plants. As

lily shoots emerge at this time of year, this is especially important when the weather is wet. After a rainy spell, it's time to restock the bait. Deadline and other liquid baits may not be as appealing to pets as pellets. Some fans have found success using beer traps, which include placing a shallow dish of beer near lily plants in the early evening.

DISEASES

Basal Rot, *Fusarium oxysporum* var. *lilii*

Symptoms

» Basal rot is characterised by a premature yellowing of the leaf, stunting, and senescence in growing plants.

» These are all common responses to ethylene gas, which is released by decomposing bulb tissues.

» Many little new scale bulbs, or bulblets, develop on the severed scales of infected bulbs. But because they develop towards the contaminated end of the scale, these bulblets are easily infected themselves.

» Many times this results in the destruction of the main bulb, but the formation of several bulblets along the stem.

Survival and spread

» The fungus can live in garden soil for at least three years without a host after being released from the decaying bulbs.

» The fungus spreads rapidly via the soil or on the surfaces of bulbs, tools, agricultural equipment, or shipping boxes.

Management

» Do not use ammonium salts or other nitrogen-rich fertilisers on bulbs since they encourage soft, quick growth that leaves the bulbs susceptible to infection.

» Mulch with compost and manure for further protection against *Fusarium*, and for the added bonus of a cooler soil environment.

» Soil moisture must be managed. Fusarium thrives in damp, low areas, therefore it's best to avoid those places while planting lilies. Be careful not to overwater plants during the hot summer months.

» When caring for lilies, it's important to keep biting insects like grubs and nematodes under control and to avoid causing any mechanical harm. Basal rot germs are able to enter through any kind of wound.

» An application of lime to increase pH may be necessary in some soils because acid soils may potentially exacerbate the condition.

» Bulb infant mortality is increased by Fusarium infection due to the high temperatures required for scale propagation. Therefore, it is essential to only scale the healthiest bulbs, to wash them completely before scaling, and to dip the scales in a fungicide before incubation.

» Commercial lily hybridizers need to put in more time and energy into breeding *Fusarium*-resistant variants of their products. This is becoming increasingly important as many of the most effective chemical controls become unavailable.

Leaf blight, *Botrytis elliptica, B. cinerea*

Symptoms

» The leaves' upper surfaces start to develop white patches (tear drop-shaped marks).

» Lighter on the edges, darker in the middle.

» The spots gradually join together, and the entire leaf wilts and falls apart.

» Severe cases show infection into the stem, leading to the death of the plant.

» In some cases, the decimation of vegetation can occur quite quickly.

» When Botrytis and humidity combine, open blooms get brown patches.

Favorable conditions

» Extreme assaults are more common in damp, humid, and muggy weather.

» Temperatures as low as minus 2 °C and as high as plus 24 °C are both conducive to infection.

» The ideal conditions for the disease are warm temperatures with persistent moisture on the foliage after periods of rain, showers, fog, or dew.

Survival and spread

» The fungus survives the winter as tiny black sclerotia on the leaves. Spores are produced by them, and they are spread to the fresh spring growth of foliage by the wind and rain.

Management

- Pull the old stems when they fall loose and dispose of them immediately once the growth season has ended.
- It's a great time to put down some new mulch.
- At the end of the growing season, many farmers treat the soil with a copper fungicide to prevent rot.
- It is best to pick off affected leaves first thing in the morning, when they are still damp, to prevent or slow the spread of the disease.
- The leaves must be dry for spraying to be effective, so that's all. Because infection typically occurs on leaf undersides, here is where focus should be.
- While Asians are very vulnerable, Orientals and Aurelians are only slightly so. The lily of the valley (*Lilium lankongense*) and its offspring have showed very high levels of resilience. The novel tetraploid hybrids appear to be more resistant to Botrytis and leaf scorch due to their thicker epidermis.

Root rot, *Rhizoctonia solani*

Symptoms

- Soil-dwelling rhizoctonia fungus is a benign lily bulb parasite.
- Spotting little lesions surrounded by dark yellow discolorations is the telltale sign that fungal mycelia are to blame.
- Numerous lesions give the scales a yellowish cast.
- Minor damage to the scales, but the sores can allow other diseases in. Infectious diseases can develop in a wide variety of strains and cause significant damage to lilies.

Favorable conditions

- Warm temperatures, such as those found in greenhouses, are ideal for rhizoctonia growth.
- When grow in soils with too fine of a texture, like heavy clays, water can't drain as easily and roots can't get enough oxygen.
- Roots become susceptible to rots after being injured by root lesion nematodes (*Pratylenchus penetrans*).

Management

- Improving cultural practises, especially drainage, is the most effective method of preventing root rots.
- A white colour returns to the bulbs if they are dipped once a year in a solution containing the chemical Quintozene, which prevents the growth of *Rhizoctonia*.

Southern wilt, *Sclerotium rolfsii* var. *delphinii*

Symptoms

- Dutch irises are particularly vulnerable to southern wilt disease.
- Like the seeds of a turnip or cabbage, the sclerotia that this fungus produces are spherical and brown or reddish in colour.
- During the hot summer months, the disease manifests itself by attacking the bulbs of actively growing plants. When temperatures rise above a certain point, sclerotia begin to germinate and begin attacking plant tissues.
- White chalky or light brown rot appears on infected lily bulbs, and the mycelia (white filaments) of the invading fungus are easy to spot.
- The disease can be identified by the presence of white, fan-shaped regions of coarse mycelia.
- Spots of brownness on plants are a telltale sign that a commercial planting row has been affected.

Survival and Spread

- The sclerotia are common in the soil near diseased plants that have died and can live for up to ten years without a host.
- Under optimal conditions, the disease can spread rapidly, rendering all of the bulbs inedible by harvest time.
- To prevent the disease from spreading, a fungicide solution can be watered around affected regions. This action is taken when localised infection is found in commercial plantings.

Management

- Raise the infected bulbs and wash them well.
- The contaminated bulbs and at least 30 centimetres of soil around them must be dug up and destroyed.

- The use of quintozene as a regulatory agent has proven successful.

Damping-off, *Pythium, Phytophthora, Rhizoctonia* spp.

Symptoms

- The first real leaf, known as the cotyledon, falls over and dies.
- Close to the ground is where the fungus strikes most.
- The seedlings don't make it.

Management

- When starting plants inside, use sterilised soil or a soilless mix.
- When starting plants from seed, only use new, sterile containers. To do this, fill a container with a 10% bleach solution and let it sit for a few minutes (sodium hypochlorite).
- When starting seedlings, use soil mixes that drain quickly and use water very sparingly.
- It's important to let the soil dry out on the surface. Watering plants early in the day allows the soil to dry a little bit before nighttime, which is important because damping-off thrives in high humidity and wet soil.
- Plants need enough of light and air to grow properly. In mild weather with plenty of light, damping-off is rarely an issue.
- Methyl thiophanate spray.

Viral diseases

- Lily Symptomless Virus (LSV), Tulip Breaking Virus (TBV), and Cucumber Mosaic Virus are three of the most prevalent viruses that infect lilies (CMV).

Symptoms

- The Cucumber Mosaic Virus can generate abnormally deformed, brittle leaves and flowers, as well as highly streaked leaf colour. Very pale leaves and stunted growth are possible symptoms. It affects a wide variety of garden plants besides only tulips, dahlias, and delphiniums.
- The fractured colour of tulips caused by the tulip breaking virus formerly captivated Dutch tulip growers some centuries ago. Mottled leaves, less vibrant dark blooms, and even colour breakdown might result from this disease in lilies.

» Symptomless lily virus: if plants aren't as perky as their neighbours, might be dealing with this sneaky disease. They are rapidly declining. Make sure the difference between plants that are suffering from the drought and those that are suffering from other causes, such as overcrowding, competition from nearby plants, damage from rabbits or slugs, or distress from the lack of water.

Transmission

» Aphids that have swallowed the virus through plant bites are a common vector for spreading it to other plants.

» When they land on a healthy plant and bite into it, they spread the virus.

Management

» Make an effort to reduce the amount of aphids that feed on lilies.

» Virus-infected plants should be disposed of immediately. Those conditions will never improve. Do not utilise them in a compost pile or the garden; instead, destroy them in a safe location.

» Disinfecting equipment used for cutting or pruning lilies or their bulbs is essential if have a virus problem. Use a solution of 1 part bleach to 9 parts water.

» It is possible to grow new plants devoid of the virus by using tissue culture and cuttings that are only a few millimetres long from the virus-free tips (meristem).

References

Kim, J., Lee, S., Kim, H., Choi, S., & Roh, M. S. (1994, August). Survey on virus diseases of Lilium spp. and their indexing by tissue blot and dot immunobinding assays. In International Symposium on the Genus Lilium 414 (pp. 189-194).

Kaminska, M., & Korbin, M. (2000, May). Detection of phytoplasma infection in Lilium sp. plants. In X International Symposium on Virus Diseases of Ornamental Plants 568 (pp. 227-236).

Nesi, B., Lazzereschi, S., Pecchioli, S., Grassotti, A., Rizzo, D., Stefani, L., & Paoli, M. (2011). Biotechnological approaches for identification of the main viruses in Lilium. Acta Horticulturae, (900), 339-345.

Lauterer, P. (2002). a New Pest of Ornamental Horticulture in the Czech Republic. Plant Protection Science, 38(4), 145-148.

10 Marigold

INSECT PESTS

Aphid, *Aphis gossypii*

Damage symptoms

- » Aphids are primarily found at the flower's growing tip, bottom surface, and petal base.
- » Both nymphs and adults feed on the sap of plants and flowers, causing them to wilt and turn a darker hue.
- » There is something distinctly unwell about the plants.

Management

- » If necessary, use a 1% spray of Neem or Pongamia oil.
- » Inject either 25 EC of oxademeton methyl or 30 EC of dimethoate at 2 ml/L into a spray bottle and mist the area.
- » Spread Imidacloprid 0.25 percent spray two or three times

Bud borer, *Helicoverpa armigera*

Damage symptoms

- » The moth lays its eggs on the buds of new plants.
- » The newly hatched larvae will dig into flower buds and developing blooms, destroying the delicate florets and petals as they feed.
- » The larvae bore into the centre receptacle of the flower, damaging it, and fed on the filaments of the ovaries and stamens, causing the blooms to wither.

» Caterpillar faeces can be spotted on the ruined blossoms.

Management

» Larvae are collected and killed once they reach maturity.

» At the first sign of egg laying on young, vulnerable foliage, spraying with a mixture of Endosulfan 30 EC @ 2 ml/L, Methyl parathion 50 EC @ 1 ml/L, or Fenvalerate 20 EC @ 0.5 ml/L and Diflubenzuron 25 WP @ 2 g/L is an effective method of preventing borer damage.

» Spray 0.75 ml/L of spinosad.

» Neem kernel extract 4% spray.

» HaNPV spraying at 250 LE/ha.

» Infected *Helicoverpa armigera* are plagued by the fungus *Nomuraea rileyi*.

Mealy bug, *Drosicha* spp.

Damage symptoms

» The young stem, leaves, and shoots are infested with mealy bugs.

» leaves that flatten and wrinkle and are a dark green colour.

» Leaves are transformed into black sooty mould due to the presence of honey-like compounds that are isolated from one another.

» The tips of the shoots are not developing as quickly as the rest.

Management

» Spray with 2 ml/L of profenophos or dimethoate.

» A 25 g/L rosin soap fish oil spray

Thrips, *Thrips tabaci*

Damage symptom

» Thrips can be identified by the discoloration or distortion of plant tissue.

» Thrips are attracted to fresh, new growth and feed mostly on it.

» When leaves are injured, they dry out and warp. The tips of infested plants may turn brown, curl, and lose their leaves before they should.

» When thrips fed on flower buds before they opened, it sometimes caused "colour break" in the petals.

Management

- Stick 20 yellow sticky traps per acre.
- Placing *Amphelicies cucumeris* spiders back into the wild.
- Azardiractin 3 ml/L or Spiromesifen 0.75 ml/L or Fipronil 1.5 ml/L.

Two spotted spider mite, *Tetranychus urticae*

Damage symptoms

- Marigolds typically suffer mite infestations in the months of August and September.
- The undersides of leaves are home to mite colonies hidden by filthy webs.
- The sap is a food source for both the nymph and the adult phases.
- Discolored and falling leaves are a symptom of the disease.
- If the attack is bad enough, the plants will die or at least stop growing.

Management

- Severely affected plant sections should be cut off and burned.
- To remove the webs off the plant, shower it well with water.
- Use Wettable Sulfur 80 WP at 3 g/L, Dicofol 18.5 EC at 2.5 ml/L, Ethion 50 EC at 1 ml/L, or Profenophos 50 EC at 1 ml/L in a spray bottle.
- Fenazaquin 2 ml/L or Abamectin 0.5 ml/L, both with a 1.9 EC.
- Pongamia oil 1% spray should be used after the aforementioned sprays.

DISEASES

Damping-off, *Rhizoctonia solani*, *Pythium* sp.

Symptoms

- Pre-emergence mortality is shown as brown necrotic spots that encircle the radical and spread downward to the plumule.
- The post-emergence symptoms of this disease manifest as a water-soaked, brown, necrotic ring on the lower half of the hypocotyls, causing the seedlings to die.
- Ground-level brown ring appears on the stem.
- Roots start to decay.

- The roots of diseased seedlings often look rotten when they are removed.

Management

- The nursery bed needs to have good drainage.
- A crop rotation cycle of three to four years is recommended.
- Carbendazim (0.1%) soil drenching is recommended for disease control.
- For efficient disease management, dilute 2 g/L of Captan or Mancozeb in water and spray it over the nursery beds.

Powdery mildew, *Leveillula taurica* (*Oidium* sp.)

Symptoms

- Tiny, white spots emerge on the surface of the leaves, and eventually they spread until the entire plant's aerial portions are covered in a whitish powder.
- The disease is characterised by a white powdery material that covers the leaves.

Management

- Once every two weeks, spray the plant with Karathane @ 0.05% or Sulfex (3g/l of water) to keep the disease at bay.
- The sickness can be controlled with a spray of Wettable Sulfur at a concentration of 3 g/L in water.

Botrytis blight, *Botrytis cinerea*

Symptoms

- Fungal growth occurs on plant tissue that is dead, dying, or damaged.
- The blossoms wilt and turn a papery brown, and grey, fuzzy masses blanket their surfaces.
- Flowers that are on their last legs are especially vulnerable.
- Tan or brown dots, resembling targets, can also emerge on the leaves.
- Flowers that have fallen to the leaf surface are a common cause of these spots.
- When a flower's components go necrotic, they eventually perish.
- Wet conditions cause necrotic tissue to get covered in a grey mass of spores.

Favorable conditions

- When prolonged periods of overcast, humid, rainy weather occur, this disease becomes very problematic.

Spread

- Pathogen sporulation and dissemination are promoted in conditions of high moisture and humidity.

Management

- Pick off and dispose of any wilted flowers or other plant components lying around your plants.
- Should not water plants directly overhead or keep them soggy for long periods of time. When watering the flowers, be careful not to soak them.
- The use of plants in confined spaces has been shown to improve air quality.
- Time for irrigation should be set for first thing in the morning.
- Thiophanate-methyl, iprodione, and chlorothalonil spray.
- Use 3.5 ounces per five gallons of water while spraying Prestop (*Gliocladium catenulatum* strain J1446).

Leaf and flower blight, *Alternaria dianthi, A. diathicola*

Symptoms

- Small purple patches spread and develop into bigger lesions with a purple border, a yellow-green border, and a gray-brown centre covered in black spores.
- This is because many lesions might join together to generate larger, uneven necrotic regions that kill the entire leaf.
- A dark brown powdery spore growth can be seen on these lesions.
- Most infections on branches occur at their nodes and their base.
- Cankers develop from these infection foci and girdle the stem, causing the branch to droop and the girdled part to become yellow and die.
- Flowers aren't safe from the fungus's wrath, however.

Favorable conditions

- This disease tends to spread during long wet periods characterised by mild nighttime showers.

Spread

» The spores travel with the wind and the rain.

Management

» Removal of agricultural debris from the planting area or timely integration of refuse after harvest are examples of cultural management strategies. Crop rotation to non- susceptible crops for two years is another option.

» It's best to not water the lawn with a sprinkler gun.

References

Wang, K. H., Hooks, C., & Ploeg, A. (2007). Protecting crops from nematode pests: using marigold as an alternative to chemical nematicides.

Raj, S. K., Snehi, S. K., Khan, M. S., Singh, M., Chaturvedi, Y., Tiwari, A. K., & Rao, G. P. (2011). Molecular detection and identification of phytoplasmas associated with little leaf and witches' broom disease of marigold (Tagetes erecta L.) in India. Phytopathogenic Mollicutes, 1(1), 41-46.

Dhiman, J. S., & Arora, J. S. (1990). Occurrence of leaf spot and flower blight of marigold (Tagetes erecta L.) in Punjab. Journal of Research-Punjab Agricultural University, 27(2), 231-236.

Sultana, R., Akanda, A. M., Haque, M. A. S., Majumdar, A., & Munsur, M. A. Z. A. (2014). An investigation to virus like diseases of marigold. J. Biosci. Agric. Res, 2, 23-35.

Gurjar, P., Meena, L., & Verma, A. K. (2019). Diseases of marigold (Tagetes erecta) and their management: a review. Int. J. Adv. Sci. Res., 4(4), 137-144.

11 Orchids

INSECT PESTS

Scale insect, *Diaspis boisduvalli*

Damage symptoms

- » Scale is flat, spherical, and yellow.
- » The nymphs and adults feed on the sap of leaves, petioles, and shoots, which can cause stunted growth and even the drying of shoots in extreme cases of attack.
- » It's harmful to foliage and 'fake' bulbs alike.
- » The growth of sooty mould is facilitated by the secretion of honey dew.
- » The leaves are starting to turn yellow.
- » Become weak and dry out.

Management

- » The use of plant material that has not been infested by pests.
- » The infected tissue is removed and burned.
- » At a concentration of 15 g/L, fish oil rosin soap is sprayed.
- » Either spray Pongamia or Neem oil at a rate of 10 ml per litre (ml/L) after applying Malathion 50 EC at 1 millilitre and Dimethoate 30 EC at 1 ml (ml/L) or Chlorpyriphos 20 EC at 2 ml (ml/L).
- » Methyl parathion 50 EC, Phosalone 35 EC, or Methyl demeton 25 EC sprayed at 1 ml/L.

- Five-seven milligm of carbofuran per litre of water.
- Larvae of the green lacewing, *Chrysoperla carnea*, were released into the wild.
- Protect predacious ladybird species like the *Cryptolaemus montrouzieri* and the *Scymnus*.

Mealy bug, *Pseudococcus maritimus*

Damage symptoms

- Both the stem and the leaf are covered in mealy bugs.
- Mealy bug nymphs and adults feed on plant sap and exude honeydew, which encourages the growth of sooty mould and attracts ants.
- Weakens plants that get it.
- Chlorotic spots caused by severe infestations can darken, eventually causing the entire leaf to become yellow and fall off before its time.

Management

- Materials for planting that have not been infested with mealybugs were chosen.
- Immediately isolate and dispose of any infected materials.
- Ants crawling around on plants, spraying for mealy bugs.
- Spot any problems and remove potential hiding places by clearing away dead leaves and flower husks.
- Pongamia oil and Neem oil sprays at 10 ml/L concentration are beneficial.
- After 15 days, spray with Methyl parathion 50 EC at 1 ml/L, followed by Dichlorvos 76 EC at 1 ml/L, or Methomyl 40 SP at 2 ml/L, or Acephate 75 SP at 1.5 g/L.
- Use 1 ml/L of Malathion 50 EC or 0.5 ml/L of Phosphamidon 85 WSC with water and spray the area.

Whiteflies, *Aleurothrix* sp.

Damage symptoms

- Whiteflies are tiny insects that look like moths and feed on fresh vegetation like flower buds.
- An infestation of whiteflies can be identified by the cloud of small white insects that rise from a plant when it is moved or disturbed.

Management

- Preventing orchid infection requires maintaining clean conditions, getting rid of weeds, and isolating orchids from their host plants.
- Malathion, Orthene, and Safer Soap are all viable options for spraying plants, provided they are used in accordance with the product labels' specific recommendations. To completely remove whiteflies, keep applying the solution every four days.

Thrips, *Anaphorathrips orchidii*

Damage symptoms

- Thrips are tiny sucking insects that feed primarily on flowers but sometimes on leaves and can spread viruses from one plant to another.
- Buds infected with pests may not open, and any blooms they do produce may be misshapen and have wet areas.
- Pitted, stippled, silvery, or bleached leaf appearances are all possible.
- Adults and juveniles alike feed on the cell sap of new growth in plants.
- Curling dark stripes appear on affected leaves.
- Discoloration, deformation, and drying of emerging buds are the results of damage.

Management

- Separating orchids from potential host plants (flowers, citrus, gardenias, eucalyptus, etc.) and using good sanitation practises are all important measures to take to avoid an infestation.
- When an infestation is discovered, spray the area with either oxademeton methyl 25 EC or dimethoate 30 EC at a concentration of 2 ml per litre.
- Imidacloprid 200 SL 0.4 ml/L, Pongamia or Neem oil 10 ml/L, or Cartap hydrochloride 50 SP 1.5 g/L if the infestation is severe.
- Pesticides such as Orthene, Malathion, and Safer Soap can be used to spray plants and flowers.

Aphid, *Macrosiphum luteum*

Damage symptoms

- Aphids are sucking insects that feed on plant tissue and spread viral diseasees from one plant to another. They are particularly destructive to new growths such as buds and flowers.

» Flowers and buds may fail to open, and a sticky residue may form on the leaves.

» Aphid honeydew attracts ants and provides the perfect environment for the growth of sooty mould.

» Check plants for aphids, mealybugs, scale, and mites if notice sooty mould.

Management

» Remove aphids from the plant by spraying it with water.

» Malathion, Orthene, and insecticidal soaps (1% Neem or Pongamia) are some examples of pesticides that can be used to spray plants.

Spider mite, *Brevipalpus* sp.

Damage symptoms

» Mites are tiny, reddish-brown pests that can be seen feeding on the underside of plants.

» Mite faeces can cause webbing and brown spots on the undersides of leaves. An injured leaf's upper surface may have a silvery sheen, but it will soon become sunken and turn dark if left untreated.

» When mites feed, the affected area becomes mottled and weak.

» Lack of chlorophyll can cause abnormal markings on leaves, such as stripes, stipples, or spots.

» As the problem worsens, the affected leaves dry out and turn brown.

Favorable conditions

» Warm, dry weather a new generation emerges every six days.

Management

» Insect infestations can be avoided by lowering temperatures and, if possible, increasing humidity and leaf moisture.

» Spraying them with horticultural oil, neem oil, mineral oil, or insecticidal soap.

» Using a spray, apply Dicofol 18.5 EC @ 2.5 ml/L, Triazophos 40 EC @ 1.5 ml/L, or Ethion 50 EC @ 1 ml, and then apply Pongamia or Neem oil @10 ml/L after 15 days.

- Abamectin 1.9 EC at 0.5 ml/L or Fenazaquine 10 EC at 1 ml/L should be sprayed when infestations are severe.
- It is important to spray plants with a miticide such as Dicofol (Kelthane), being sure to get the undersides of all the leaves. There will have to be multiple applications, maybe even three at 4-day intervals.
- Greenhouses benefit greatly from the usage of the predatory mite *Phytoseiulus persimilis*.

Snail, *Achatina fullica*

Damage symptoms

- Snails can damage plants by eating the leaves, blooms, and even the roots.
- At night, both adults and juveniles feed on vegetation.
- The monsoon season is when snail attacks are most common due to the high humidity levels.
- As a means of digestion, they scrape the leaves, creating large, uneven holes, and also snip the young, fragile shoots.

Management

- Snails are collected by hand and killed in a 5% salt solution.
- Spreading crystallised salt along walkways or around planters.
- Snail Kill (pellets containing 3% metaldehyde) @ 50 kg/ha should be spread.
- To treat plants, spray them with Neem oil (at a rate of 10 ml/l) or soapnut extract (at a rate of 60 g/L).
- It is possible to use chemical baits in the garden. Spreading ash or diatomaceous earth on flat surfaces can act as a barrier, but it will be rendered ineffective by moisture.
- Spreading beer in shallow tins in the growing area will kill the pests and they can be easily removed the following day.

DISEASES

Black rot, *Pythium* and *Phytophthora* spp.

Symptoms

- Small, uneven, watery, brown dots appear on the underside of leaves first and then quickly darken to a purple brown or black.

- The edges of the spots may be becoming yellow as they spread.
- The lesions become bigger over time and may leak fluid.
- Sometimes older lesions dry out and turn black, inviting additional disease attacks.
- High temperatures and humidity may accelerate the disease's spread to the rhizome and roots.
- Simply remove fresh leads, which have a purple or purple-brown centre and a yellowish extending edge.
- Infections in pseudo bulbs, roots, or rhizomes manifest as a purple-black, often clearly demarcated discoloured patch at the plant's heart.
- It's not uncommon for the infection to begin in the plant's roots and work its way up to the base of the pseudo bulb or leaf, eventually killing the plant.
- Some cattleyas will develop a yellowish tinge on one or both sides of the pseudo bulbs.
- There will be a change in hue to a dark brown or black when the discolouration diminishes and the bulb rots.

Favorable conditions

- Humidity and heat both play a role in the rapid proliferation of the disease.

Management

- To prevent the spread of the disease from one plant to another via splashing water, it is best to remove the affected plant entirely.
- After removing contaminated tissue using a sterile instrument, a fungicide such as Subdue or Banrot should be applied.

Fusarium wilt, *Fusarium proliferatum*

Symptoms

- By clogging the phloem, Fusarium prevents water from entering the plant's tissues.
- Yellowing, thinning, shrivelling, wrinkling or wilting, and ultimately death, characterise infected leaves.
- Upon slicing the rhizome, a purple or pinkish-purple discoloration will appear in a circle or band on the outer layers.

- All of the rhizome and even the pseudo bulbs could turn purple if the disease is severe enough.
- Plants with a severe infection may die in 3-9 weeks, while those with a minor infection may slowly wither away over the course of a year.

Spread

- The fungus is transmitted from plant to plant as a result of insufficient hygiene, most commonly through the use of non-sterile cutting equipment.

Management

- By taking care to practise good hygiene, avoid contracting the disease.
- If a purple ring can be seen, the infected rhizome or pseudobulb must be thrown away.
- Douse the sterilised plants with methyl thiophanate.
- Be sure to thoroughly clean the cutting area and any instruments.
- When the blade comes into touch with contaminated tissue, it must be disinfected before being used again, ideally by flame.

Root rot, *Rhizoctonia* sp.

Symptoms

- *Rhizoctonia* is most prevalent in the plant's roots, although it can also affect the plant's stems and leaves.
- New growths are getting smaller while older leaves and pseudo bulbs are turning yellow, shrinking, becoming thin, and becoming twisted.
- Brown rot and white or brown fungal growth on the roots are common symptoms.
- When the fungus invades a plant, it girdles it and eventually destroys it.
- Infected spores spread rapidly through the rhizomes and lower leaves of young plants.
- Brown root rot quickly kills infected plants.

Favorable conditions

- Root rot develops in the event that the medium degrades, drainage is inadequate, or the plants are overwatered.

» Roots that have been injured or have accumulated salt from hard water or excessive fertilisation quickly rot.

Management

» Use a clean knife to cut away diseased roots and leaves.

» Follow the label directions for applying a systemic fungicide (like Subdue) or a protectant fungicide (like Thiophanate methyl) to the surviving plant.

» Use a bleach solution of 10% to sterilise the planting area.

» Verify that the potting mix is new and that the plants are not drowning in their own water.

Anthracnose, *Colletotrichum* and *Glomerella* spp.

Symptoms

» Most attacks occur on the leaves.

» Beginning at the leaf 's tip and working its way down, the leaf 's colour gradually changes to brown.

» Sometimes they appear as concentric rings, while other times there are multiple dark bands throughout the leaf 's surface.

» Typically, the damaged area is clearly demarcated and slightly sunken, while the rest of the leaf appears unaffected.

» Spore bodies form in the affected area.

» Flowers acquire watery, black or brown pustules which are usually elevated and occur on the underside of older sepals and petals.

» It's possible that the spots will join forces and spread across the entire flower.

Favorable conditions

» When temperatures are high and light levels are low, the pathogen thrives.

Management

» The spread of the disease could be slowed by implementing standard hygiene measures, improving ventilation, decreasing temperatures (if at all possible), and increasing lighting.

» Thiophanate methyl, a systemic fungicide, or Mancozeb, a protectant fungicide, may be sprayed. Switch between using a protectant fungicide and a systemic one.

Petal blight, *Botrytis* sp.

Symptoms

» Flowers get tiny black or light brown patches.

» Possible expansion of the spots to encompass the entire flower.

» The presence of moisture can cause a grey fungal growth on badly infected or rotting flowers.

Management

» Sanitation measures may be effective in lowering infection rates.

» Boost the flow of air.

» Reduce humidity.

» Make sure the lows are above 68 degrees at night.

» Flowers that have been contaminated should be disposed of immediately.

» Protective fungicides, such as Daconil, should be sprayed in closed spaces.

Leaf spots, *Cercospora*, *Septoria*, *Guignardia/Phyllosticta* spp.

Symptoms

Cercospora

» One of the earliest symptoms of infection is a yellow spot on the underside of a leaf.

» On the upper side of the leaf, the yellow-green patch.

» The dots spread and grow in a radial or asymmetrical manner, possibly covering the entire leaf.

» Eventually, the ageing spots will turn a dark purple colour, necrotic and sunken.

» The leading edge of the advancement is still yellow.

» Extremely infected leaves often drop off the plant before their time, especially if the disease manifested itself at the leaf's root.

» The top surface of the leaf first gets chlorotic and then necrotic.

Guignardia/Phyllosticta

» Tiny, yellow, slightly sunken lesions appear on the leaf or pseudo bulb.

- Particularly if the disease has spread to the leaves, they will take on an oval shape and become depressed as they grow larger.
- As they get older, they go from light tan to dark brown and gain a somewhat ruffled, reddish-purple-black rim.
- Spores, which are microscopic black structures, become higher in the middle of the spots over time.
- Early leaf drop can occur in severely diseased plants.

Septoria

- Tiny patches that are initially sunken and yellow in colour might appear on any leaf surface.
- They continue to grow and eventually turn a very dark colour, often bordering on black, and take the shape of either a circle or an irregular lesion.
- It's possible for spots to join together to produce larger, erratic regions on the leaf.
- Heavy infection causes early leaf drop.

Management

- Provide proper sanitation with good air flow.
- Water on the leaves might encourage infection, so try to keep them as dry as possible.
- A sterile instrument should be used to remove diseased leaves, and leaf moisture should be decreased.
- Spray with a protectant fungicide, such as Mancozeb, or a systemic fungicide, such as Thiophanate methyl. To prevent resistance, switch between systemic and protectant fungicides.

Bacterial soft rot, *Erwinia* spp.

Symptoms

- On the leaves, little water spots emerge, typically encircled by golden halos.
- The infection will spread quickly through the leaves and roots, and then more slowly through the rhizomes or pseudo bulbs if it is allowed to progress unchecked.
- This wet rot looks and smells bad because it is soaked in water.

- Plants in the Phalaenopsis family can become fully rotten in as little as two to three days if the disease spreads aggressively.
- Dendrobium leaves turn black and droopy when they are saturated in water, appearing yellow at first.
- Translucent spots on vanda leaves get black and sunken when the plant is stressed.
- Small, circular spots appear on paphiopedilum leaves, starting off yellow and waterlogged before turning reddish brown and sunken.
- Before the spot reaches the leaf tip, it spreads out in all directions and may even reach the expanding crown.
- The plant will quickly shrivel up and turn black if the disease is allowed to grow unchecked.
- The grammatophyllum leaves develop water-soaked brown patches that eventually turn black and sunken.

Favorable conditions

- If the disease is present, watering from above could transmit it to other areas.
- Hot and damp environments are ideal for the pathogen.

Management

- Use a sterile instrument to immediately remove the contaminated tissue.
- Remove moisture from the air, dry the leaves, and lower the temperature and humidity (if possible).
- The affected plants and their surroundings should be sprayed with a bactericide such as Physan or Copper compounds.
- Use a bleach solution of 10% to sterilise the planting area.
- In order to keep infections at bay, it is recommended to use copper compound sprays on a regular basis, especially in hot and humid climates.

Bacterial brown spot, *Acidovorax* sp. (syn. *Pseudomonas* sp.)

Symptoms

- The symptoms manifest as tiny, tender, water-filled blisters on the leaf anyplace.
- The diseased area, which was initially a nasty green colour, swells, coalesces, and finally turns brown or black, dried out, and sunken.

- When the disease progresses to the leaf's tip, it causes the leaf to leak a liquid that is high in germs.
- Older Cattleya plants are more susceptible to infection because of the wounds they have had in the past. It's sluggish to spread and almost seldom lethal.
- These blister-like patches on Phalaenopsis may have a yellow or pale green halo. When spots join together, the disease spreads swiftly. The plant will perish if the sick spot spreads to the crown.

Favorable conditions

- During the warmer months, its prevalence increases.
- *Pseudomonas cattleya* is a bacterium that spreads through water and thrives in humid, warm environments.

Management

- Bring down the relative humidity and temperature (if possible).
- Do away with watering from above.
- Boost the flow of air.
- Any contaminated tissue must be removed immediately.
- Use a bleach solution of 10% to sterilise the planting area.
- Bactericide spray, such as Physan or Copper compounds, should be applied to the plant.

Viral diseases (*Cymbidium Mosaic Virus, Tobacco Mosaic Virus, Tobacco Spotted Wilt Virus, Orchid Fleck Virus, Tospovirus*)

Symptoms

- Chlorotic and necrotic patches, streaks, lines, and rings can be seen on the leaves as symptoms.
- Flowers may get colour fading, necrotic patches, and streaks.
- As a systemic pathogen, the viruses can be found in every organ of the plant.

Management

- Remove the plant is the only way to ensure no other plants are harmed by it.
- To prevent the spread of viruses from plant to plant via the sap, only use clean cutting implements.

- » Always use latex gloves when working with plants, and throw away the gloves once done. The virus can be spread from plant to plant bare hands come into touch with infected plant sap.
- » Maintain a sanitary potting surface. When repotting, keep a stack of newspapers close by, and use them to line the bottom of the container. Be sure to discard the newspaper, gloves, and any other trash accumulated.
- » Verify the cleanliness of the containers.
- » Remove each inflorescence off the plant with a clean instrument.
- » If a virus is present, it could be transferred to hand without knowledge, allowing to infect the next plant touch or the next dead flower bud you remove.

References

Kumari, S., & Lyla, K. R. (2006). A survey of the pests of orchids. Journal of Tropical Agriculture, 39(1), 32-34.

Watson, J. B. (2002). Orchid pests and diseases. American Orchid Society, Inc..

Ting, W. P., & Ho, T. H. (1970). Diseases and pests of orchids. Bulletin, Ministry of Agriculture and Lands., (124).

Meena, N. K., Medhi, R. P., & Mani, M. (2016). Orchids. Mealybugs and their Management in Agricultural and Horticultural crops, 525-534.

Meena, N. K., & Mani, M. (2022). Pests and Their Management in Orchids. Trends in Horticultural Entomology, 1239-1254.

Sangma, R. H. C., & Singh, D. R. (2020). Pests of Orchids and Their Management. In Advances in Pest Management in Commercial Flowers (pp. 63-84). Apple Academic Press.

12

Rose

INSECT PESTS

Red scale, *Aonidiella urantii*

Damage symptoms

- The severity of the scales is at its peak just before and after the monsoon.
- Yellowing, leaf drop, and branch and twig death are all symptoms of an infestation on the upper surfaces of plants.
- Adult and juvenile scales steal the life force from healthy stems and branches.
- Dead or severely damaged shoots will be encrusted with a reddish brown colour and will have dried out.
- The stem is entirely covered by scales and the affected area of the plant dries out.
- Whole plants can rot from the inside out if they get infected by servers.

Pest identification

- Nymphs are less than 0.5 mm in length and have a pale yellow, oval shape.
- The female's crimson scales are shaped like a circle and measure around 2.5 mm in diameter with a raised dot in the middle. The bug is yellow under the scale, which is a reddish brown colour. Females are big and flightless.

Management

- Selecting plants that haven't been exposed to scaling.

- Disposal of infected plant material via cutting and burning.
- Use cotton dipped in kerosene or diesel to remove the scales.
- Apply a spray containing either Malathion 50 EC at 1 ml/L, Chloropyriphos 20 EC at 2.5 ml/L, Dimethoate 30 EC at 2 ml/L, or Ethion 50 EC at 1 ml/L.
- Use 1.0 kg a.i./ha (33 kg/ha) of Carbofuran 3 G after pruning.
- Spray 10% Pongamia oil on newly trimmed branches.
- Natural predators including *Aphytis melinus*, *A. lingnanensis* (around the coast), and *Comperiella bifasciata* are released.

Thrips, *Rhipiphorothrips cruentatus, Scirtothrips dorsalis*

Damage symptoms

- Deformities in flowers, leaves, stems, and shoots are caused by insects that feed by sucking the plant's sap.
- Because of their small size, thrips are largely invisible unless look closely at the undersides of flowers or leaves (which become silvery on the upper surface).
- When female thrips lay their eggs within the tissue of a leaf, the resulting offspring will do damage to the plant by rasping the tissue and drinking the cell sap that flows out.
- A thrips infestation follows any sort of trimming, and the resulting new growth is immediately attacked.
- Nits and adults alike feed on the cell sap of young plants.
- Most commonly, damage manifests as curled leaves with brown spots and malformed buds with scorched margins.
- Flowers with pest infestations don't bloom properly or at all, and their petals dry out and fall off too soon.

Pest identification

- *R. cruentatus* - The adult *R. cruentatus* is a dark brown, while the juveniles are a bright red.
- *T. tabaci* - Both nymphs and adults have a very light, almost white, golden hue.

Management

- » The act of loosening the soil in preparation for winter.
- » Two to three applications of 0.5% Pongamia oil sprayed with Acephate 75 SP at 1.5 g/L or Dimethoate 30 EC at 2 ml/L should be done every 15 days, coinciding with the emergence of a new flush.
- » In cases of severe infestations, Fipronil 5 SC @ 1.5 ml/L should be used.
- » Chlorpyrifos 20 EC @ 5 ml/L should be sprayed heavily on the ground to kill pupae.
- » Spraying 0.5% Pongamia oil when first signs of disease appear.
- » Pests can be controlled by spraying a mixture of Vertimac (a biopesticide made from the bacterium *Streptomyces avermitilis*), 4 percent NSKE, 2 percent Neem oil, and 1 percent Jatropha oil or Annona oil.
- » Spraying 0.05% Quinalphos or Phosphamidon and applying 0.18 g a.i. of Carbofuran to the soil per plant should be done as well.

Aphids, *Macrosiphum rosae, Aphis gossypii, Myzus persicae*

Damage symptoms

- » Aphids, which are a blackish green tint, congregate in groups near new shoots and buds.
- » Adults and juveniles alike sap plant vitality by sucking the contents of plant cells.
- » Honeydew is a pleasant sticky material that wasps and ants excrete, and it also promotes the formation of sooty mould.
- » Rapid aphid development is encouraged by the high humidity and overcast weather that typically occurs between October and November.
- » The drying up of young shoots.
- » Leaves become misshapen and bud development is slowed.
- » Flowers begin to show signs of wilting and lost buds at an early stage.

Pest identification

- » *M. rosae:* Pear-shaped body that is lengthy and pink in hue; has enormous red eyes and black cornicles.
- » *A. gossypii:* Round and a range of colours from brown to black to yellowish green.

- *M. persicae:* Colors of yellow-green or, occasionally, red.

Management

- Apply Neem oil, Pongamia oil, or Dimethoate 30 EC at a concentration of 10 ml per litre or 2 ml per litre, respectively, at the first sign of infestation.
- The use of Imidacloprid 200 SL at 0.4 ml/L or Cartap hydrochloride 50 SP at 1 g/L as a spray should be considered for serious outbreaks.
- Liminool 0.1% or Methyl parathion 0.25 % spray.
- Aphids can be prey for lady beetles, *Coccinella septempunctata*, and *Chrysopa oculata.*
- Use 3g/L of sprayed *Verticillium lecanii.*

Bud borer, *Helicoverpa armigera*

Damage symptoms

- At get to the petals, larvae drill holes into the buds.
- When feeding, the animal will insert only its head into the bud, leaving the rest of its body exposed.
- Damage to flowers occurs during their mature stages.
- Petals of several of the flowers have holes.

Pest identification

- Larvae are of a sage green colour and have dark grey lines running laterally across their bodies.
- In its adult form, this moth has a black mark on its forewings and a V-shaped speck and a dark border on its hind wings.

Management

- Borer populations can be reduced by collecting and destroying adult larvae.
- If eggs are found on young, fragile foliage, spray a mixture of Diflubenzuron 25 WP at 2 g/L and either Chlorpyriphos 2 ml/L, Methyl parathion 50 EC at 1 ml/L, or Fenvalerate 20 EC at 0.5 ml/L.
- Effectively controlling the bud borer population at 1.29 and 1.57 larvae/plant and a net BCR of 6.00 and 8.34, respectively, were applications of Indoxacarb 0.0075% and Spinosad 0.002%.

» Use a 4% Neem seed kernel extract spray or a 1% Neem oil spray.

» HaNPV spraying at 250 LE/ha or *Bacillus thuringiensis* formulation at 1 kg/ha.

» It is recommended to use 1,000,000 *Trichogramma chilonis* or 500,000 *Chrysoperla carnea* per hectare.

Whitefly, *Trialeurodes vaporariorum*

Damage symptoms

» Leaf withering, chlorosis, leaf drop, and/or reduced growth can all be directly attributed to whitefly infestation.

» They also cause "honeydew," which serves as a food source for other bugs and a breeding ground for the "sooty mould" fungus.

Pest identification

» The adults are about the size of a pinhead (1/16-inch-long), with powdery white wings and short antennae that make them look like moths.

» Nymphs without wings tend to have a scaly, flattened shape.

Management

» Put out sticky yellow traps, traps with no bait, trap lights, or pheromones to lure them in.

» For this, diseases such as *Beauveria bassiana, Paecilomyces fumosoroseus*, and *Verticillium lecanii* are utilized.

» Utilize parasitoids such as the *Encarsia formosa* and the *Eretmocerus eremicus.*

» Use predators such as the Miridae- *Macrolophus pygmaeus*, the coccinellidae, *Delphastus pusillus*, and the Phytoseiidae- *Amblyseius swirskii.*

» Tobacco plant extracts, Neem oil, horticultural oils, pyrethrums, and garlic water in spray form.

Budworm, *Spodoptera litura*

Damage symptoms

» As their name suggests, early instars are social insects that feed by scraping chlorophyll from plants.

» Mature, black-colored larvae cause significant harm to flower buds and stems at night, reducing floral quality.

Pest identification

- The length of an adult moth is around 20 mm (0.59 inches), while its wingspan is around 30 mm (1.18 inches) (1.18-1.5 inches). The front wings range in colour from grey to a rusty brown, and they feature a complicated pattern of white veins and creamy streaks. The edges of the back wings are a drab greyish brown.
- Mature brown larvae appear to have three narrow yellow longitudinal lines, one on the top or dorsal side and one on each lateral side. Along the lateral lines is a row of black dots, and along the dorsal line is a row of dark triangles.

Management

- Egg masses and gregarious first-instar larvae are collected and disposed of.
- Using a sprayer, apply either Quinalphos 25 EC at 2 ml/L or Endosulfan 35 EC at 2 ml/L every 15 days.
- If the frequency is very high, spray Indoxacarb 14.5 SC at 1 ml/L or Thiodicarb 75 WP at 1 g/L.
- Apply 250 litres of Sl NPV per hectare, followed by 1-2 ml/L of Neem formulations.
- To remove the adult larvae, set up a poisonous trap using wheat/rice bran, jaggery, and Chlorpyriphos (10:1:0.5).

Two spotted spider mite, *Tetranychus urticae*

Damage symptoms

- Smaller, cream-colored nymphs and larger, red adults with two black dots have been spotted scraping and sucking.
- Eat the undersides of leaves and spin silken webs to trap prey.
- Affected leaves become discoloured, bronzed, and dry as a result.
- Tender buds also show discolouration, which slows growth and dries it off.
- In the event of a serious infestation, the population will spread to all parts of the plant's foliage, including the flowers and buds.
- Feeding causes the upper surface to develop yellow dots and the leaves to turn crimson.
- Finally, the affected leaves die off.

- The effects on development and blooming are negative.

Pest identification

- White, spherical eggs are placed on the underside of the leaves.

Management

- Damaged plant parts are cut off and burned if the infestation is too severe to be ignored.
- Reduce heat and pest populations through good air circulation, pest-free farming practises, and regular watering.
- Spraying with Dicofol 18.5 EC at 2.5 ml/L, Wettable Sulfur 80 WP at 3 g/L (Pirone, 1978), 0.05% Oxydemeton methyl or 0.012% Fluvalinate (Pokharkar et al., 1986), Profenofos 50 EC at 1 ml/L, or Ethion 50 EC at 1 ml/L, followed by Pongamia, Neem, very effective new miticides are available, such as Vertimec (0.025%), Flufenoxuron (0.01%), Diafenthiuron (0.025%), Profenphos (0.5%), and Amitraz (0.5%).
- To prevent a proliferation of mites and to maintain better control in a polyhouse setting, sprays containing 0.5 to 1% Neem or Pongamia oil should be applied in the evening and before the commencement of summer.
- Mite population is successfully controlled after release of predatory *Phytoseilus persimilis* at a density of 26/ m^2.

DISEASES

Black spot, *Marssonina rosae* (syn. *Diplocarpon roseae*)

Symptoms

- Coal-black leaf patches with black edges and yellow halos are diagnostic of this disease.
- Both sides of a leaf will get spots.
- Older leaves in the bottom part of the plant are the first to show symptoms, leading to significant defoliation and a decrease in both yield and bloom size.
- When viewed through a hand lens, these patches reveal the tiny black blister-like fruiting bodies of the fungus.
- Consistently stripping a plant of its leaves can lead to its eventual collapse, death, or at least a significant reduction in its strength and vitality.

- Stems of particularly vulnerable species may show a drastic weakening of plant vitality.

Survival and spread

- The fungus survives the winter on dead rose leaves and canes.
- Wind and water splashing from rain or other sources are the primary vectors for the dissemination of the fungus spores.

Management

- Plant roses in a spot where the morning sun will swiftly evaporate the dew.
- In order to improve airflow, it is recommended to implement cultural practises such as increasing distance between plants and thinning the canopy's centre growth.
- Keep the foliage as dry and avoid watering it from above.
- Sanitation entails the elimination of disease by means of the elimination of contaminated canes and the burning of unhealthy leaves.
- Remove any damaged or diseased stems in the spring before new leaves grow.
- Rose black spot can be prevented with the fungicides Tebuconazole in combination with Trifloxystrobin or Triticonazole, as indicated on their product labels.
- Plants of resistant types (such as 'Pride N Joy' hybrid tea, 'Sexy Rexy' floribunda, and 'Prima Donna' grandiflora) should be used.
- Fungicides and fungal biocontrol agents (such as *Trichoderma harzianum* and *Chaetomium globosum*) (Chlorothalonil, Mancozeb).

Die-back/stem canker, *Leptosphaeria coniothyrium*, *Botryodiplodia theobromae*

Symptoms

- Springtime browning and shedding of shoot tips (usually the result of frost damage).
- Pruning stubs turning brown and drooping as they die back, which eventually affects the rest of the branch (the result of a poor pruning technique).
- A seasonal or annual occurrence characterised by the death of individual plant parts or the entire plant (various causes).

» Due to drying, cracks can occur in these locations, and the resulting sores can girdle the stems and ultimately kill the plant.

» Small black fruiting bodies, which are fungal formations, can sometimes be seen on the damaged plant portions.

» Root rot can occur concurrently in some cases.

Favorable conditions

» Wet spells, high humidity, close spacing of vegetation, and diminished airflow.

Spread and survival

» Dead branches and flower stems are still home to the fungus.

» Stem canker and die-back fungi are notoriously hardy, and can often overwinter on damaged canes or plant waste.

» Typically, spores of the causative fungi are carried by wind and rain or irrigation water.

Management

» Since wounds are a key method the fungus enters the plant, it is important to take care not to injure the plant when transplanting, cultivating, trimming, or cutting flowers.

» Roses should be planted in loose soil that has been thoroughly amended.

» By performing this pruning, we can prevent the development of lesions on the young shoots.

» Dead or damaged wood should be removed immediately.

» During this time, also remove any weak or crossing branches.

» When pruning, make sure cuts are smooth and made directly above an outward-facing bud, sloping away from it.

» To the trimmed off ends, apply either 0.2% Carbendazim, Benomyl, or Copper oxychloride, or 0.2% Chaubatia paste, Bordeaux paste, or Rose paint (Copper carbonate, red lead, and linseed oil in a 4: 4: 5 ratio).

» Spray with 0.2 percent Copper oxychloride, 0.2 percent Difolatan, 0.2 percent Chlorothalonil, or 0.2 percent Mancozeb.

» A Manetti rootstock may do slightly better against typical dieback.

Rust, *Phragmidum mucronatum*

Symptoms

- Yellow pustules, indicating damage, emerge on the undersides of the leaves and stems. After then, a dark red hue appears.
- Damaged leaves become misshapen and yellow, then fall off unexpectedly.
- The orange papules develop into black papules by late summer.
- A die-back symptom may also show up if the plant's health is declining.
- The leaves are falling off too soon.

Survival and spread

- The pustules that initially produced the lighter, more fragile summer spores begin to generate the darker, more robust resting spores toward the end of summer.
- In many cases, these spores spend the winter stuck to stems or trellises.
- Typically, spores of the causative fungi are carried by wind and rain or irrigation water.

Management

- Make sure there's plenty of ventilation.
- Avoid planting roses in tight spaces, and remember to prune so that the plants always have room to grow.
- Be sure to water plants before noon, and never soak the leaves.
- Raking and incinerating of leaf litter.
- To stop the disease spores from being splashed back up onto the leaves, cover the soil with a thick layer of mulch or organic compost.
- Apply a spray of Carboxin 0.1%, Wettable Sulfur 0.25 %, Captan 0.2 %, Mancozeb, Lime sulphur, Triforine, or Zineb.
- *Bacillus subtilis* has proven to be an effective treatment for rust disease.

Anthracnose, *Sphaceloma rosarum*

Symptoms

- Some people mistake anthracnose, which is caused by the fungus *Sphaceloma rosarum*, for black spot on rose leaves.

- Spots caused by anthracnose are distinct from other types because they are smaller and have more distinct borders at first.
- The plant starts getting these sunken greyish-brown patches.
- The centres of the spots turn grey and may fall out as they grow larger, giving the appearance of shot holes.
- Within a matter of days, affected leaves wilt and fall off, actively expanding shoots may become brown, and developing buds dry up and fall off.
- Reddening and yellowing of the leaves and even their premature fall are possible symptoms, though not to the same extent as with black spot.

Survival and spread

- There, the spores can survive the winter in the wood.
- Fungus spores can travel on wind currents or in raindrops.
- Both natural precipitation and a sprinkler system can be used to disseminate them around the garden.

Management

- Do away with the infected stuff and throw it away.
- Cut rose bushes back so there is space in the middle for air to circulate.
- Don't water plants from above, and cutting back on excess foliage. This aids in the rapid drying of the plant's foliage and helps avoid the growth of fungi in damp, dark places.
- Apply a spray containing Chlorothalonil, Mancozeb, Triforine, or Zineb on the shrubs.
- Plant hardy Rugosa roses like "Blanc Double de Coubert," "Frau Dagmar Hartopp," "Will Alderman," and "Harison's Yellow Rose (Rosa x harisonii)" to withstand the elements.

Downy mildew, *Peronospora sparsa*

Symptoms

- A rose plant infected with the devastating downy mildew disease can lose all of its leaves in a matter of days.
- The leaves and stems develop purple spots bordered by yellow. These spots typically appear on the veins.

- Underneath the leaves, a white, downy fungus can develop.
- After a while, the leaves will dry out and fall off.

Favorable conditions

- Downy mildew is triggered by wet weather, high relative humidity, and cloudy skies.
- The circumstances for an epidemic are greatly improved by fog.
- A temperature of around 18 °C is ideal for the growth of downy mildew.

Survival and spread

- Rose bushes are a good overwintering location for spores (leaves and canes).
- Since the downy mildew pathogen can hibernate as a specialised spore (oospore) in infected rose debris and soil, it can continue to exist between rose crops even when exposed to extreme weather.
- Rose downy mildew oospores can spread via wind or water.

Management

- Downy mildew can only be managed by using both cultural and fungicidal methods.
- Humidity needs to be lowered first.
- Greenhouses need adequate ventilation to thrive.
- Plants in a greenhouse or nursery should be spaced far enough apart to allow for adequate air circulation and drying of their leaves.
- Drip irrigation is another method for minimising leaf moisture.
- The infection can persist on detritus for several weeks, so growers should remove fallen leaves and prune away infected plant sections.
- Aliette, Banol, Segway, Stature, and Subdue MAXX are just a few of the fungicides approved for use in commercial greenhouses and nurseries.
- Copper + Mancozeb is an option.

Powdery mildew, *Sphaerotheca pannosa*

Symptoms

- Young, vulnerable leaves, shoots, and buds show the symptom as greyish-white powdery spots.

- Younger leaves are more likely to be purplish and to curl over, revealing the underside, if they are affected by the fungus; these leaves may also have elevated blister-like patches that are coated with the fungus's white powdery development.
- Malformed adult leaves and unopened flower buds are symptoms of infection, and overall, plants look wilted.
- Disease can spread from the tips of young canes and destroy them if it does.
- There is a change in petal colour, shrunken size, and eventual death.
- Buds infected with a disease may not open, and those that do may only bloom poorly.

Favorable conditions

- Conditions favourable to the growth of powdery mildew on roses are warm, dry days followed by cool, damp evenings.
- When daytime temperatures are between 15 and 20 °C and night time temperatures are between 20 and 27 °C with high relative humidity, it can happen virtually anytime during the growing season.
- It's particularly bad in the shade and throughout the winter.

Survival and spread

- This type of fungus survives the winter as mycelium in dead but not completely died buds and shoots.
- Conidia or ascospores are the primary inoculum of fungi.
- Conidia carried on the wind account for the secondary spread.

Management

- Raking and incinerating of leaf litter.
- When plants are dormant is the best time to prune their stems.
- Mancozeb, Sulfex, Wet-sulf, or Wettasul at 0.2% or Benomyl, Carbendazim, Triademefon, Fenarimol, Penconazole, or Triforine at 0.1% should be sprayed on the plant's foliage two to three times every 15 days during the growing season.
- Chlorothalonil, Propiconazole, Piperalin, Fenarimol, Copper hydroxide, Neem oil, Azoxystrobin, Kersoxim methyl, Ziram, and Triforine may be used.

- 25 kg/ha of sulphur dust (Use of Sulfur at higher temperature conditions will be phytotoxic).
- Apply a spore suspension containing *Ampelomyces quisqualis* and *Pseudozyma flocculosa* to the plant's foliage to prevent further infestation.
- Utilizing a concentration of 3.5 g/liter of K2HPO4 or KH2PO4.

Botrytis blight/Gray mold, *Botrytis cinerea*

Symptoms

- Botrytis cinerea, often known as grey mould, causes a fuzzy gray-brown growth on rose blossoms and buds.
- The fungus thrives in humid environments between 17 and 22°C.
- Discolored sunken regions (cankers) and dieback, which may extend down the stem from the flowers, are symptoms of infection in canes.
- Small, white spots on the petals of infected flowers are surrounded by reddish halos and can rapidly enlarge into big, erratic dots.
- Sometimes infected buds will not open at all, will droop, and will develop sunken, smooth, greyish-black sores.
- It's possible for a mycelial growth that is greyish brown in colour to spread throughout the entire bud and even into the stem.
- The tips of the petals may turn dark, and the petals may also develop a wide variety of little brown specks, circular spots, or blister-like areas.
- Cane blights, also known as cankers, form on infected stubs or pruning wounds.
- Girdling canes can lead to stem failure and cane death.

Favorable conditions

- Climates that are chilly, rainy, and humid are ideal for the spread of disease.
- The pathogen *B. cinerea* requires a temperature of 15°C in order to thrive.

Survival and spread

- Infected plants have a special overwintering structure called sclerotia.

Management

- Eliminate the sick plant and all of its surrounding detritus and tissues.

- » Dry out greenhouse by turning up the heat, opening some vents, and/or cranking up the fans.
- » Put plants in strategic locations to increase ventilation.
- » Replace overhead sprinklers with a drip system.
- » Top-dress or mix calcium nitrate, gypsum, or another suitable calcium source into the growing medium in the container.
- » Use Thiophanate methyl, Chlorothalonil, or Neem oil fungicides and spray the area.
- » Copper fungicides can be applied to trees and bushes when they are dormant.
- » Bio-agents such as *Trichoderma harzianum*, *Trichoderma viride*, and *Pseudomonas flourescens* should be sprayed.

Crown gall, *Agrobacterium tumefaciens*

Symptoms

- » Galls found on stems or at the soil's surface are rough and spherical in shape.
- » At the stem's base, spherical outgrowths develop, each measuring about 5 cm in diameter.
- » Galls vary in appearance according on their age, with young ones being white or pale green and older ones being dark and woody.
- » Galls weaken and stunt plants by blocking the movement of water and nutrients from the roots and stems to the plant's upper parts.
- » Plants that succumb to this condition often lose their vitality and develop unusually small leaves and fewer blooms than normal.
- » The damaged plants quickly wilt and stunted growth.

Survival and spread

- » The soil-dwelling bacteria have a potential 3-year lifespan if no plants are present.
- » Water currents carry the bacteria, which enter the rose plant via grafting or cultivation wounds or rodents and insects.
- » Infectious bacteria can be transferred from one garden to another via implements.

Management

- Choose disease-free roses to avoid problems with crown gall.
- Once a plant is infected, nothing can be done since there are no chemical controls available for crown gall.
- Since the bacteria enter through fresh wounds, it is important to avoid damaging the plant's roots and crown as much as possible during planting and cultivation.
- As soon as galls are spotted, afflected plants should be thrown away.
- It is best to dig up the sick plant and dispose of the soil it was growing in.
- In order to prevent crown gall, it is necessary to disinfect any tools that have come into contact with it.
- Tools can be disinfected by soaking them in a solution of 0.5% sodium hypochlorite for a few minutes (household bleach).
- Successful management of crown gall has been achieved through the use of a commercially available formulation of a non-tumerogenic strain of *Agrobacterium rhizogenes* (the K-84 strain).

Mosaic disease, *Rose Mosaic virus* (RMV)

Symptoms

- Yellow line patterns on the leaves are a symptom of this virulent disease, which is caused by a trio of viruses (*Prunus necrotic ringspot virus* (PNRSV), *Apple mosaic virus* (ApMV), and *Arabis mosaic virus* (ArMV).
- Chlorotic bands in an oak-leaf pattern, chlorotic ring spots, small blotches, and line patterns in waves are only a few examples of the wide variety of designs that can be found.
- Some or all of the leaves may display the patterns.
- This virus infects plants when they are grown from cuttings taken from infected plants.
- Plants that are infected weaken and are more susceptible to damage from other stresses such as drought or cold temperatures.
- Yellow spots on foliage and misshapen shoots are symptoms of Rose Mosaic Virus.
- It may be a benign infection or it may restrict growth.

» Weakened strength.
» Having a reduced life expectancy.
» Lower levels of flowering.
» Flowers on shorter stems.
» The leaves have become misshapen, puckered, or shrunken.
» Very low rate of success after transplant.

Survival and spread

» Disease is typically spread by grafting infected tissue or buds onto infected rootstock.
» Vegetative proliferation from infected buds, scions, or rootstocks is the only known route of transfer.

Management

» Strict adherence to hygiene standards in nurseries, the generation of disease-free planting material, and the elimination and destruction of all contaminated plants all contribute to the containment of the blight.
» Plants infected with viruses are beyond rescue.
» In established rose gardens, rose mosaic spreads very slowly through root grafts, if at all.
» Remove infected plants from valuable gardens and dispose of them properly.
» Stay away from those that exhibit signs of mosaic disease.

Rosette disease, *Rose Rosette Virus* (RRV)

Symptoms

» The clumping of the stems, like a witch's broom, is another characteristic.
» Reddish-orange leaves and stems on stout, elongated canes.
» Extremely sharp, tiny thorns of red or brown tint.
» Abnormal or nonexistent flowering.
» Leaves that are either underdeveloped or abnormally thin.
» Maybe some bent canes.
» Canes that have died, leaves that have turned yellow or brown.

- Having a short stature or appearing to have been stunted.
- Assorted of the preceding.

Spread

- During feeding, the rose leaf curl mite spreads and introduces it into the rose (*Phyllocoptes fructiplilus*).
- Infected nursery material can also be disseminated through grafting, budding, and propagation techniques.
- Diseases like rose rosette can be passed from one rose to another if they are planted too closely.

Management

- Keep roses at least a foot apart before planting.
- Plant only seedlings guaranteed to be free of Rose Rosette Virus.
- Plants that have been contaminated should be wiped out.
- Any multiflora roses within a 100-foot radius of the intended roses should be removed.
- Because the sap on pruners used on infected plants is contaminated with the virus, they must be sterilised with rubbing alcohol or a mild bleach solution before being used on healthy plants.
- Spraying neighbouring roses with Bifenthrin every two weeks from April to September will help prevent the spread of eriophyid mites from an infected plant.

References

Horst, R. K., & Cloyd, R. A. (2007). Compendium of rose diseases and pests. APS press.

Bout, A., Boll, R., Mailleret, L., & Poncet, C. (2010). Realistic global scouting for pests and diseases on cut rose crops. Journal of economic entomology, 103(6), 2242-2248.

Debener, T., & Byrne, D. H. (2014). Disease resistance breeding in rose: current status and potential of biotechnological tools. Plant Science, 228, 107-117.

Bout, A., Boll, R., & Poncet, C. (2009). Quick pest and disease scouting to implement IPM in greenhouse rose crops. IOBC/WPRS Bull, 49, 155-160.

Handayati, W., & Sihombing, D. (2017). Recent Status of Pests and Diseases on Cut Roses in Batu East Java. KnE Life Sciences, 639-647.

Hegde, J. N., Ashrith, K. N., Suma, G. S., Chakravarthy, A. K., & Gopalkrishna, H. R. (2020). Insect pests of roses and their management. Advances in pest management in commercial flowers, 85-101.

Arya, S. (2018). Insect Pest and Diseases of Roses: Symptoms and Control Measures. Academic Social Research:(P),(E) ISSN: 2456-2645, Impact Factor: 5.128, Peer-Reviewed, International Refereed Journal, 4(1).

13 Aloe

INSECT PESTS

Aphid, *Aloephagus myersi*

Damage symptoms

» Adults and juveniles alike will forage for food near the leaf bases or in the rolled tips of broken leaves.

» Moreover, their honeydew production is fueling the spread of black mould.

» Stunting and growth retardation occur as a result of a severe infestation.

Management

» Use of insecticidal soap and the protection of natural enemies are two examples of environmentally friendly pest management strategies.

DISEASES

Rust, *Phakopsora pachyrhizi*

Symptoms

» Any opening in the plant's defences, such as a cut or scrape, might serve as an entry point for fungi.

» Small, light yellow or brown spots appear on leaves, grow, and eventually turn brown due to the disease.

» The rust causes lesions of varying shades of orange and brown to appear on the undersides of leaves.

» When a surface is infected with fungus, it will not spread to other areas.

» The plant could lose its leaves.

Management

- If the problem is serious, spray a fungicide on the leaves and remove the ones that are infected.
- In many cases, rust infestations can be avoided by simply keeping plants in a sunny location with adequate air circulation.
- Remove dirt and debris from the soil and any cracks or crevices in the plant.
- Take care not to over-water the plant while it is infected.
- Apply water directly to the soil using a drip irrigation system.
- Fungicides available for purchase could potentially be useful in finding a solution.

Leaf spot, *Alternaria alternata*

Symptoms

- There were large necrotic deep circular or oval dark brown stains on both sides of the leaves.
- Impacts not only the plant's development and productivity but also the quality of its end product (bad odour of the gel).

Spread

- Ascospores are the primary vector of infection.
- Secondary dispersal by airborne conidia.

Management

- Using a spray of Mancozeb or Chlorothalanil at a concentration of 0.2% is an excellent method for preventing leaf spot disease.
- Even when exposed to a high concentration (3.1 104 conidia/ ml) of pathogenic spores, two plant growth boosting rhizobacterial strains, namely *Burkholderia cenocepacia* VBC7 and *Pseudomonas poae* VBK1, were able to significantly lower disease severity.

Soft rot, *Pectobacterium* (*Erwinia*) *chrysanthemi*

Symptoms

- The outside petiole, which is in direct touch with the ground, is the first to become infected and subsequently spreads throughout the plant.

- Infected leaves are moist and typically exhibit full-on leaf rot as a result of the infection.
- The infected area darkens in colour and softens, becoming mushy.
- Infected leaves rot slowly and leak water as they disintegrate.
- Internal rot in infected leaves causes wilting in young leaves, which is followed by tip collapse and death of the entire plant.

Favorable conditions

- Contaminated plant waste that was abandoned in the field to decay.
- Injuries and wounds in plants.
- The sickness is brought on by hot, wet weather with lots of rain.

Survival and spread

- Diseases thrive in hot and damp conditions, and bacteria can survive in plant detritus in the field.
- The bacteria are carried both initially and subsequently by water.

Management

- When weeding, take care not to damage the plants.
- Infected aloe plants should be rooted out and thrown away.
- Soybeans with forage legumes as part of a crop rotation.
- Applying 400 kg per hectare of defatted Neem cake to the soil before planting and immersing seedlings in 0.3% Copper oxychloride 50 w/p solution + 0.05% Streptocycline (Streptomycin sulphate + Chlortetracycline hydrochloride) significantly decreased disease incidence.

References

Ayodele, S. M., & Ilondu, E. M. (2008). Fungi associated with base rot disease of Aloe vera (Aloe barbadensis). African Journal of biotechnology, 7(24).

Avasthi, S., Gautam, A. K., & Bhadauria, R. (2015). Occurrence of leaf spot diseases on Aloe vera (L.) Burm. f. caused by Curvularia species from Madhya Pradesh, India. Biodiversitas Journal of Biological Diversity, 16(1).

Singh, N., Iqbal, Z., Ansari, T. A., Khan, M. A., Ali, N., Khan, A., & Singh, M. (2019). The portent plant with a purpose: Aloe vera. Journal of Pharmacognosy and Phytochemistry, 8(3), 4124-4130.

14 Asparagus

INSECT PESTS

Asparagus beetles, *Crioceris asparagi*, *Crioceris duodecimpunctata*

Damage symptoms

» The asparagus beetle causes harm to the plant by munching on the tips of new growth.

» As soon as the leaves emerge, asparagus beetles and their larvae begin gnawing on the surface of the stems and devouring the new growth.

» If the fern is severely damaged, the crown will be weakened, which is especially problematic in a young asparagus stand.

Pest identification

» The full-grown larvae of the asparagus beetle (*Crioceris asparagi*) are around 9 mm long with a dark greenish grey colour. A blue-black body and a red prothorax characterise the adult beetle. Their elytra, or wing covers, are yellow with red borders.

» The adults of the spotted asparagus beetle, *Crioceris duodecimpunctata*, are a ruddy orange colour with six large black dots on each wing cover.

Management

» Spinetoram (Radiant) SC should not be used on ferns within 60 days of spear harvest, and should be used only after harvest.

» At the fern stage, use Chlorpyrifos (Lorsban Advanced); only two applications after harvest per year are allowed. We must prevent any surface water pollution from drift or tail water.

» Methomyl (Lannate) LV can be used as needed every 7 days. The maximum rate per acre of crop application is 4.5 lb a.i.

» Ferns and brush growth benefit from an application of Carbaryl. If necessary, reapply once every 7 days.

» For best results, use Ambush, Pounce exclusively with ground forces and hardware. A maximum of 0.4 lb a.i./acre each growing season is recommended.

Armyworms, *Spodoptera praefica*, *S. exigua*

Damage symptoms

» The asparagus fern caterpillar is nocturnal and feeds mostly on the branch epidermis during night.

» Constantly encircling the branch, killing the ferns within.

Pest identification

» A mature beet armyworm (*Spodoptera exigua*) looks like a little moth with spotted grey or dark wings.

» Eggs have a striated appearance and might be a pastel green or pink.

» Caterpillars reach a maximum size of about 1.25 inches in length. Their bodies can range in hue from olive green to nearly black, and they always have a bright yellow stripe running down each side.

» The adult western yellow striped armyworm (*Spodoptera praefica*) is a tiny, mottled grey or dusky-winged moth.

» The typical caterpillar is a dark colour with two wide, vivid stripes on either side. It grows to be around 2 inches long when fully mature.

Management

» Controlling weeds effectively is an important step in avoiding an armyworm infestation.

» When spraying an organically certified crop, use *Bacillus thuringiensis* or the Entrust formulation of Spinosad.

» When armyworms are young, use *Bacillus thuringiensis* subsp. *aizawai* to kill them.

» They are harmless to the environment's natural predators.

- Coragen should not be applied at a rate greater than 15.4 flu oz a.i. per acre of crop, and products containing Chlorantraniliprole should not exceed 0.2 lbs a.i. per acre of crop.
- Chlorpyrifos (Lorsban Advanced) shouldn't be used after harvest more than twice.
- We must prevent any surface water pollution from drift or tail water.
- Methomyl (Lannate) LV can be used as needed every 7 days. Use no more than 4.5 lb a.i./acre/crop.

Cutworms, *Peridroma saucia*, *Euxoa messoria*

Damage symptoms

- Each night, larvae devour tiny holes in the tips of fresh asparagus spears for a meal.
- Curling of the spears may also be caused by only feeding one side.
- The variegated cutworm eats both subterranean and aboveground plant life.
- Injuries to ferns are unusual.

Pest identification

- The mature larva of the variegated cutworm (*Peridroma saucia*) is about 3 centimetres in length, and it is yellow to brown in colour with a row of 4 to 6 yellow or pink diamond-shaped spots down its back.
- The adult dark-sided cutworm (*Euxoa messoria*) is about 5 cm long, with uneven, longitudinal stripes, and has a dark grey to greenish grey tint.
- The wings of an adult cutworm moth are dark grey or brown in colour, and their front wings often have irregular patches or bands.

Management

- Effective weed control in and around the field, as well as effective incorporation of field debris and previous crop residues, can help prevent egg and worm overwintering.
- Dig about 2.5 centimetres into the soil surrounding a damaged spear to check for signs of damage and confirm the presence of cutworm.
- If pests return in devastating numbers, use 5% Carbaryl (Sevin) bait and reapply 7-14 days later. Apply no more than 20 lb of spears per acre while spreading them on the ground.

- For variegated cutworm, a low dose of Methomyl (Lannate) LV should be used. Use as needed, every 5 days to a week. Use no more than 4.5 lb a.i./acre/crop.
- Use only ground-based equipment to apply Permethrin (Ambush, Pounce). A maximum of 0.4 lb a.i./acre each growing season is recommended.

Aphid, *Brachycorynella asparagi*

Damage symptoms

- Growth and branching are skewed as a result of being fed (broomed appearance).
- Aphid-infested shoots continue to bloom long after the rest of the field has finished blossoming.
- Damages plants by stunting their root development and, in extreme cases, killing off seedlings and young plants.

Pest identification

- Both the adults and the juveniles are a powdered bluish-gray colour with a milky, opaque consistency. As its name suggests, these aphids are tiny, measuring only approximately 0.04-0.0625 inches in length, and they also lack any obvious cornicles, which are paired tubes that extend from the back of the abdomen to the sides.
- There may be wingless and winged varieties growing on the plants.

Management

- Grasp the asparagus and pull any stray plants.
- After harvest, check for aphids by beating plants against a white surface once or twice a week. Try this out in a variety of spots and on different plants across the field.
- Douse the fern phase with Chlorpyrifos. Prevent any surface water pollution from drift or tail water.
- After the asparagus harvest is complete, spray the remaining ferns with Pyromezine (Fulfill).
- Start treating at the first sign of insect infestation with a solution of Pyrethrin (PyGanic) 1.4 EC in water that is sufficient for complete covering. Application methods using air blast are superior to those using concentrate.

Asparagus miner, *Ophiomyia simplex*

Damage symptoms

- Newly planted fields are under attack.
- The nectar and plant fluids provide the adults with their food source.
- Discolored reddish-brown tunnel in a serpentine pattern along the base of the stem is a classic sign of mining damage.

Pest identification

- It's a little black fly, hardly more than 0.2 inches in length.
- The length of the brownish red pupae is roughly 0.2 inches.
- Maggots can be white and eel-like (in their early instar stages) or round and plump (later stages).

Management

- Use foliar applications of insecticide when there are the most adults around.
- Over the winter, clear the fields of any debris to prevent insect buildup, and either destroy the material or relocate it far away from the fields where asparagus is grown.
- The asparagus fields need to have the invasive plants removed.

DISEASES

Crown and spear rot, *Phytophthora asparugi*

Symptoms

- The soft, water-soaked lesions caused by *Phytophthora* spear rot can appear on shoots at or near the soil line.
- Rapid elongation and light browning of the lesions is observed.
- The spear's damaged side will get flattened as the lesion shrinks and collapses, and the shoot will become very bent and might break.
- Young infected storage roots have the appearance of being drenched in water but yet being rigid.
- Infected crowns with *Phytophthora* spp. exhibit a yellowish-orange hue to their tissue. Damaged tissue looks soggy and fibrous when an infection is severe.

Survival and spread

- In extremely damp conditions, the soil-borne fungus *Phytophthora* attacks the shoot at or near the soil line.
- Heavy spring rains might generate major disease losses.
- If spears are hydrocooled before being shipped to the market, the disease could spread to other spears and cause significant rot.

Management

- Use disease-free transplants and *Phytophthora*-free soil as much as feasible.
- Don't overwater and make sure there's adequate drainage.
- If symptoms occur, treatment may be necessary.
- The antibiotic Mefenoxam (Ridomil Gold SL) should be applied 30-60 days before the first cutting. To ensure maximum success, a second treatment should be made soon before harvesting begins.
- Apply after planting seedlings or after covering 1-year-old crowns.
- Apply Fosetyl-AL (Aliette) once over the top to fully developed asparagus ferns.

Crown and root rot, *Fusarium oxysporum* f. sp. *asparagi*

Symptoms

- *Fusarium* causes a slow but steady loss of productivity and growth in mature plants.
- During the summer, infected plants are distinguished by one to several stunted, bright yellow ferns.
- A reddish brown vascular discoloration, which may extend into the crown, is present at the base of stalks infected by *F. oxysporum* f. sp. *asparagi*.
- Reddish specks or sunken brown lesions can be found when the crowns and underground regions of stems are cut apart.
- Reddish brown, circular lesions form on store roots of infected plants.
- Reddish brown discolouration appears on feeder roots, many of which may have rotted away entirely.

Survival and spread

- Since the fungus may live in the soil continuously, it could potentially spread through contaminated seeds.

- The disease can spread from one field to another through the movement of infected soil, on transplants grown in infected soil, and on plants grown from contaminated seed.

Management

- Produce disease-free seedlings, transplants, or crowns by planting sterile seed in pest-free soil.
- Long rotations out of asparagus are beneficial.
- Reduce the amount of stress on the plant.
- Increased plant vigour in hybrid types like UC 157, Apollo, and Jersey Giant confers some tolerance but not resistance to the disease.
- Avoid protracted harvest periods and halt harvest when production decreases to 70% of the season's greatest yield to maintain crown vitality. Maintain crown vitality with adequate irrigation and fertilising.

Rust, *Puccinia asparagi*

Symptoms

- Premature defoliation of the fern, stunting and dying of young plants restricts the photosynthetic power of the plant, diminishing carbohydrate reserves in the crown for the following year's production.
- Urediospores are orange while teliospores are black in appearance.
- This can affect spear quality and marketable production and make the plant more sensitive to *Fusarium* and *Phytophthora*.
- Reduces lifetime of plantings.

Management

- Scout for aeciospores on young asparagus or early volunteer plants in early summer.
- Plant crop rows with ample spacing and in the direction of the prevailing winds to enhance air movement and minimise times of prolonged foliar moisture.
- Plant asparagus cultivars that are moderately disease resistant or tolerant (slow rusting).
- Remove any asparagus that sprouts up on its own. Destroy abandoned fields.

- » Controlling rust promotes fern vitality and may benefit in managing soil-borne diseases.
- » Put fungicides to use.
- » Risking human and environmental safety by burning crop trash in late fall or winter is impractical on vast plots of land.
- » Erosion from wind, diminished spear quality and marketability, and crown and root damage are all consequences of tilling in order to bury crop trash.

Purple spot, *Stemphylium vesicarium*, *Pleospora allii*

Symptoms

- » Premature defoliation of the fern inhibits photosynthetic potential of the plant, diminishing carbohydrate reserves in the crown for the following year's production.
- » There are lesions on the stem, and they are purple.
- » This makes the plant more vulnerable to *Fusarium* and *Phytophthora*, and can lower spear quality and marketable production.
- » The lifespan of plants is shortened as a result.

Management

- » Assess the farms for any signs of the disease.
- » Fungicide applications can be planned with the help of the TOM-CAST disease forecaster.
- » Managing soil-borne pathogens may benefit from suppressing purple spot, which improves fern vitality.
- » Risking human and environmental safety by burning crop trash in late fall or winter is impractical on vast plots of land.
- » Erosion from wind, diminished spear quality and marketability, and crown and root damage are all consequences of tilling in order to bury crop trash.

References

Selvaraj, K., Sivakumar, G., Veeraraghavan, V. P., Dandannavar, V. S., Veeraraghavan, G. R., & Rengasamy, G. (2019). Asparagus Racemosus-A Review. Systematic Reviews in Pharmacy, 10(1).

Saravanan, L., & Chaudhary, V. (2012). Life history and seasonal incidence of

Asparagus beetle, Lema downsei Baly on Asparagus racemosus in India. Annals of Plant Protection Sciences, 20(1), 83-87.

Ramanna, D., Prasad, K., & Goud, K. B. (2010). Pest complex of medicinal plants. Karnataka Journal of Agricultural Sciences, 23(1), 197-199.

15 Aswagandha

INSECT PESTS

Epilachna beetle, *Henosepilachna vigntioctopunctata*

Damage symptoms

- Those cigar-shaped, yellow eggs are the beetles' offspring, and they're laid in bunches on the undersides of leaves.
- Both adults and grubs harm leaves and soft parts by scraping the epidermal layer in a netted pattern.
- In late stages of crop development, when Epilachna beetles are prevalent, the leaves become completely skeletonized.
- Damaged plants lose their leaves and become shrivelled as a result of dehydration.
- The plants dried out and died off in the end.

Pest identification

- Eggs are placed in clusters on the undersides of leaves, they are long, yellow, and spindle-shaped.
- A grub has a yellowish body, is stocky, and is covered in spines.
- The pupa is yellow and has spines down its back.
- The adults of this species are round and about medium size; they have a yellowish brown tint, and they have 14 black dots on each elytron.

Management

- Eggs, larvae, and adults are all harvested and destroyed, which drastically reduces the population.
- The beetle can be efficiently stopped from feeding by spraying 2% Neem oil.
- In the early stages of an infestation, spraying twice at 15-day intervals with 5% Neem seed kernel extract has shown to be effective.
- Malathion 1% or Fenthion 1% spraying.
- Spread 2 ml/L of either Endosulfan (35 EC) or Quinalphos (25 EC) through spraying.
- Using Carbaryl 50 WP as a spray use Wettable Sulfur 50 WP 2g/L in water.
- Three species of Eulophid parasitoids sp. (grub parasites), and *Tamarixia* sp. (egg parasitoid) - preyed on the beetle during its egg and grub stages.
- Parasitic wasps, like *Pediobius foveolatus*, can eliminate as much as 52 percent of an epilachna beetle infestation by killing off the host insects.
- Spraying a mixture of 0.2% Carbaryl (50 WP) and 0.2% Wettable Sulfur (50 WP) on collected beetles, pupae, and grubs was found to be effective.
- Neem oil, at a rate of 3 percent, was mixed with FYM at a rate of 12.5 tonnes per hectare and the commercial biofertilizer Azophos, which contains a symbiotic community of *Azospirillum* and *Phosphobacteria*, at a rate of 2 kilogm per hectare.

Cut worm, *Agrotis ipsilon*

Damage symptoms

- Cut-off seedlings and mature plants' shoots are easy prey for larvae.
- The plant's growth is stunted and its yield is reduced due to the damage.

Identification of Pest

- The larvae have a drab brown tint with a contrasting reddish head.
- An adult has a brown body and black markings on its wings. It has black, patched forewings and lighter, unmarked ailerons.

Management

- Pick out the caterpillars by hand and smash them.
- A 5% solution of NSKE should be sprayed.

- Before planting, dust with Malathion 50 EC, 0.1%.
- Apply a 2 ml/L solution of Chlorpyriphos 20 EC by spray.
- To remove the larvae, set traps with poison bait.
- Spraying 500 litres per hectare (L/ha) of a 5% Neem seed kernel extract (NSKE) solution was found to be successful in conjunction with hand-picking larvae, using poison bait to attract and kill them, and allowing time for the bait to break down.

Fruit borer, *Helicoverpa armigera*

Damage symptoms

- Worm way into the fruit and chow down on the young leaves of *W. somnifera.*
- After hatching, newborns consume young leaves.
- Developing larvae eat holes in fruit.
- It gets inside the fruit, where it feeds in secret.
- They ate holes in the developing fruit, which greatly reduced the overall harvest.

Management

- *Trichogramma chilonis* field discharge at 5 ml/ha.
- 0.5 kg of *Bacillus thuringiensis* (Bt) per hectare.
- HaNPV was sprayed at a rate of 250 LE/ha.
- Pheromone traps should be placed at a density of 12 per hectare.
- Spraying Bt at 0.5 kg/ha or *Helicoverpa* Nuclear polyhedrosis virus (NPV) at 250 larval equivalent (LE)/ha and installing pheromone traps (12 traps ha-1) were found to be successful when combined with field release of *T. chilonis* at 100,000 eggs/ha.

Oleander hawk moth, *Deilephila nerii* (*Daphnis nerii*)

Damage symptoms

- The oleander hawk moth's larvae cause significant defoliation by devouring the plant's leaves.

Pest identification

- The larva has a thick body, is greenish in colour, and has a protrusion on the dorsal side of the last abdominal segment that resembles a horn.

- The adult has a huge frame, is stocky, and is a dark green colour with white patches.
- The wings are long in front and short in back.

Management

- Pupae are vulnerable to avian predation because the soil is dug up to expose them.
- Caterpillars are hand-picked and wiped off.
- One light trap per hectare is installed.
- A trap plant, Nerium *(Nerium olender)*, is planted around the field.
- Spray a mixture of 2 ml of chlorpyriphos 25 EC and 2 ml of quinalphos per litre of water.

Mealy bug, *Ferrisia virgata*

Damage symptoms

- Remove the sap from the undersides of the leaves and the pods.
- Yellowing and drying of infected leaves is a telltale sign of a fungal infection.

Management

- It was discovered that applying FYM to the soil at a rate of 12.5 t/ha + Azophos at a rate of 2 kg/ha + Neem cake at a rate of 1 t/ha was effective, along with a foliar spray of Neem oil at a rate of 3 percent, as needed.
- Neem oil, at a rate of 3 percent, was mixed with FYM at a rate of 12.5 tonnes per hectare and the commercial biofertilizer Azophos, which contains a symbiotic community of Azospirillum and Phosphobacteria, at a rate of 2 kilogm per hectare.
- It has been discovered that *Cryptolaemus montrouzieri* grubs and adults feed on mealy insect larvae and adults.
- From November to January, there was an uptick in predator activity.

Aphid, *Lipaphis erysimi*

Damage symptoms

- The sucking action of the pests causes the lower surfaces of the leaves and fruits to become affected with sooty mould.

» They may do a lot of harm to plants and multiply quickly.

Management

» Spraying the infected area with Neem seed kernel extract (4%).

» Both *Verticillium lecanii* (2 g/l) and Garlic Chilli Kerosene Extract (GCKE) (5 %) were more effective in lowering aphid population 15 days after the third spray than Dimethoate (1.7 ml/L), Pongamia oil (3 ml/L), Neem oil (5 ml/L), NSKE (5 %), and Nimbecidine (3 ml/L).

» After being sprayed with 1% Gronim (a neem-based product), aphid populations dropped by 65%.

» With a cost-benefit (CB) ratio of 1: 16.8, two sprayings with 0.025% Oxydemeton- methyl (25 EC) to suppress *A. gossypii* boosted crop output by 40%.

» Use a foliar spray with a mixture of 0.5% Malathion and 0.1% to 0.3% Kelthane every 10 to 15 days.

» Using a spray containing 1% *Bacillus thuringiensis* (Bt), the aphid population dropped by 59%.

» The aphid population can also be reduced by predators like the *Chrysoperla carnea* and the *Cheilomenus sexmaculatus.*

» Using an integrated management strategy that included seed treatment with Neem leaf powder (1%), followed by foliar sprays of Neem leaf extract in water (10%), reduced aphid populations from 9.5 aphids/plant in the control group to 3.7 aphids/plant.

Ash weevil, *Myllocerous viridanus*

Damage symptoms

» Adult weevils are destructive because they eat leaf edges in a jagged, erratic fashion.

» Weevils, or grubs, eat plant roots, leading to plant death.

Pest identification

» Grubs are a pale greyish tint.

» The adult has a long, thin snout and is a drab ash-grey tint.

Management

- Capturing and killing off adult and juvenile stages.
- Carbaryl 50 WP at 3 g or Malathion 50 EC at 2 ml/L of water can be sprayed.
- Applying Chlorpyriphos at a rate of 0.1% to the soil.

Mite, *Tetranychus urticae*

Damage symptoms

- It spins web-like constructions on the undersides of leaves, where it feeds.
- Swarms of mites can be seen on the tops of leaves.
- After feasting on leaves, the insect leaves behind silvery white dots.
- Deplete yield by sucking the lifeblood out of leaves, which causes them to wither and turn yellow.
- It's common for leaves to gradually turn grey, white, or a pale brown during a severe assault, followed by a rapid and unexpected loss of all foliage.
- When mite populations are high, they spin a web across the entire top of the plant.

Pest identification

- The adults of this species have two black dots on their otherwise dark purple-red skin.

Favorable conditions

- They thrive during the hot and dry months of the year (April-June), when the average maximum temperature increases to 35 °C or more, and feed mostly on plants growing in direct sunlight.

Spread

- Spider web/net—like structures generated by the mites, linking the leaves of neighbouring plants, allow the pest to spread to nearby uninfected plants.

Management

- Application of Dicofol (2.5 ml/L) at 15 DAS.
- Spray the area with Ethion at a concentration of 10 ml/L.
- To eliminate *T. urticae*, spray oxademeton—methyl, which has a fast knockdown impact and a strong residual activity.

- When compared to Dimethoate (1.7 ml/L), Garlic Chilli Kerosene Extract (GCKE) (5%), NSKE (5%), Pongamia oil (3 ml/L), and Nimbecidine (3 ml/L), *Verticillium lecanii* (2 g/L) and Neem oil (5 ml/L) were found to be more effective in lowering the mite population.

DISEASES

Root rot, *Fusarium solani*

Damage symptoms

- Plant mortality due to the diseases ranged from 30-50%. (Gupta *et al.*, 2004).

Symptoms

- At an advanced stage, the disease produces significant wilting, which in turn promotes the decomposition and death of subterranean sections.
- When the soil was wet, a cottony growth surrounded the diseased roots.
- Brown and mushy, diseased roots are a sure sign of infection.
- Causing plant death and a drop in yields.
- Plants with root rot tend to be stunted in growth and vigour, with fewer and/or smaller leaves, blooms, and fruits than their healthy counterparts of the same age.
- When a plant's roots decay, it might cause a delay in flowering.
- This results in wildly varying yields.

Management

- Determine which plants have been infected with Fusarium wilt and promptly dispose of them.
- Plants should be cultivated in soil with good drainage to stop Fusarium from taking hold.
- Curing disease via flood fallowing.
- It was discovered that seeds treated with SAAF 0.2% (Carbendazim 12% + Mancozeb 63% WP) and *Trichoderma viride* (10 g/kg seed) + Soil application of Neem cake were effective.
- Make use of Carbendazim and Benomyl, two systemic fungicides.

Witches' broom, *Phytoplasma*

Symptoms

» Little leaves, phyllody, clusters of highly proliferating branches with reduced internodes, and a bushy, witches' broom appearance are all signs that follow in the wake of the initial infection.

Management

» Clones free of disease can be created using mesenchymal stem cell cultivation.

» Tetracyclines have bacteriostatic effects on phytoplasmas, meaning they impede the pathogens' development.

References

Venkatesha, M. G. (2006). Seasonal occurrence of Henosepilachna vigintioctopunctata (F.)(Coleoptera: Coccinellidae) and its parasitoid on Ashwagandha in India. Journal of Asia-Pacific Entomology, 9(3), 265-268.

Arvind, K., Singh, C. P., & Renu, P. (2009). Insect pests of Ashwagandha, Withania somnifera Linn. in tarai region of Uttarakhand. Entomon, 34(2), 115-118.

Marimuthu, T., Suganthy, M., & Nakkeeran, S. (2018). Common pests and diseases of medicinal plants and strategies to manage them. New Age Herbals: Resource, Quality and Pharmacognosy, 289-312.

MH, S. K., & Srinivas, M. P. (2018). Pests attacking medicinal and aromatic plants in India: a review. Journal of Entomology and Zoology Studie, 6(5), 201-205.

16 Coleus

INSECT PESTS

Mealy bug, *Orthezia insignis*

Damage symptoms

- Puncture the coleus and drink its fluids for sustenance.
- The roots of coleus plants are also a favourite food for several kinds of mealy bugs.
- Honeydew is a sweet secretion that insects like aphids produce.
- Females often release a cottony sticky material at stem junctions.
- The leaves on your coleus will shrink and its stems will shrivel.

Management

- Some people have had success in getting rid of them with insecticidal soap or rubbing alcohol diluted to 25% or 50% with water.
- In addition, chemical pesticides are readily available.

Tinged bug, *Monantha globulifer*

Damage symptoms

- Adults with lacey wings and black spiny nymphs both feed on leaf sap.
- In severe situations, the affected leaves will develop yellow spots, turn brown, shrivel, and die.

Pest identification

- » Those who encounter a nymph will notice her black coloration and sharp spines.
- » Adults have tiny to medium bodies and membranous wings with black dots on the forewing.

Management

- » The infected leaves should be thrown away.
- » Spray Acephate or Carbaryl 50 WP at 3 g/L in water at the first sign of an infestation.

Spike borer, *Helicoverpa armigera*

Damage symptoms

- » The larvae eat the plant's leaves and then destroy any newly formed spikes.

Pest identification

- » The larva has a bluish-green hue and dark, shattered grey streaks running along its flanks.
- » The adult moth has a stocky build and is either brown or yellowish brown in colour; it measures 20 mm in length and has a pair of large, black dots on the middle of each of its forewings. The wings' edges are black, and the wings themselves are a pale, muted tone.

Management

- » Picking off diseased spikes by hand and killing any larvae you find is the recommended course of action.
- » Use 2 ml of Endosulfan 35 EC per litre of water for spraying, or 1 ml of Parathion 50 EC.

DISEASES

Wilt, *Fusarium chlamydosporum*

Symptoms

- » A decrease of vitality and premature mortality were seen in the infected plants in the field, along with a progressive yellowing, marginal necrosis, and withering of leaves.

- Infected plants' tap and lateral roots completely rotted and turned a strange colour.
- This kind of plant had very peelable bark.
- Much of the affected bark peeled away and was shredded.
- Root and collar rot proved fatal for the damaged plants.
- Infected tubers began to decay and give off an unpleasant odour.

Management

- Due to the risk of severe *Fusarium* infections, water stagnation in *C. forskohlii* fields should be avoided.
- By using natural insecticides like Neem cake, *C. forskohlii* was able to reduce its disease index by 40-60% and boost its root yield.
- To prevent fungal diseases in *C. forskohlii* plants, dip the tips of your cuttings in a Carbendazim solution (1 g/L) before planting.
- Using emisan (0.2% of the total) has been shown to provide some protection against *Fusarium wilt* for the plants.
- Disease rates were lowered thanks to benomyl and carbendazim.
- Inhibiting *Fusarium* growth is one of *Pseudomonas fluorescens*' (Pf) RB50 and RB31's many talents (Mallesh, 2008).
- *Trichoderma viride, Glomus fasciculatum, Glomus mosseae*, and *Pseudomonas fluorescens* are all bio-inoculants that can be added to soil to boost its ability to produce forskohlin in plant roots.
- The most effective method of disease prevention was an inoculation with both *T. viride* and *G. mosseae*.
- For the lowest wilt incidence (12.76 percent), combine *T. viride* with Neemato (a neem-based treatment applied at 500 g/5m^2).

Downy mildew, *Peronospora* sp.

Symptoms

- Downy mildew causes brown discoloration on the leaves, which makes the plant look unclean. It can also cause the leaves to bend and twist.
- Brown patches or flecks appear irregularly on the leaves, and the plants may sometimes twist and lose their leaves.

» Spots on leaves sometimes appear as squares or rectangles bordered by bigger veins.

» Leaves' undersides look "filthy" because of the grayish-brown spores that have settled there.

Spread

» Diseases can easily spread when sick plants are transported or when employees bring infected plants into contact with healthy ones.

» In addition, it can be blown or splashed from one plant to another.

Management

» Infected plants must be removed from the garden.

» Mancozeb is excellent for its protective action against downy mildews, and Coppers (0.1% Copper oxychloride) are also useful.

» Chemicals like Aliette and Alude (phosphorous acid compounds) and Compass and Heritage (strobilurins) that work systemically or partially systemically are effective against downy mildews.

Leaf blight, *Rhizoctonia* spp.

Symptoms

» Infected areas first showed up as water soakage, and then the infected tissues at the plant's collar region morphed into a mushy, black, watery mass.

» The infection spread to the roots, where it caused rot and eventually the plant's demise.

» Discoloration and root hair decay were visible on infected plant roots.

» The affected bark would slough and peel off in large areas.

» This disease was found to spread swiftly in humid environments.

» If the infection is very severe, the plant will lose its leaves and eventually die.

Favorable conditions

» During the monsoon or other times of excessive humidity, blight disease tends to flourish.

Management

» *T. viride* + Neem to had the lowest percent blight.

Root rot, *Sclerotium rolfsii*

Symptoms

- The disease first showed itself as a darkening of the stem in the collar region of the plant.
- Withering and falling, the leaves grew limp.
- Small, light to dark brown sclerotia developed on the infected part of the stem as white, fan-shaped mycelial strands spread outward from the stem.
- The sclerotial letters started out white and aged to a brownish colour.
- As time progressed, the plant withered and died.

Management

- Sanitizing the field, immersing stem cuttings in 0.1% Carbendazim, and then drenching the entire plant again with 0.1% Carbendazim 30 days after planting are all part of the Carbendazim treatment regimen (DAP).
- Application of 20 kilogm of zinc sulphate per hectare plus 50 gm of a Neem cake and *Trichoderma viride* mixture each plant.

Bacterial wilt, *Ralstonia solanacearum*

Losses

- Heavy losses (>50%) are being caused by wilt in south India.

Symptoms

- Infected plants have water-soaked spots with linear streaks around the collar area.
- The plants rapidly wilted and dried out as their leaves turned limp and droopy.
- The plant's leaves began to curl and eventually die off.
- With a light tug, wilted plants came off, and discolorations were seen in the vascular system.
- There was leaking of bacterial exudates when these tubers were crushed.

Management

- *C. forskohlii* fields should be kept as dry as possible to prevent Ralstonia infections, which can be fatal.

» Streptocycline 300 ppm solution applied to the soil around the roots is effective.

» Maximum plant height, number of branches, fresh and dry weight of the plants with decreased disease severity, and root rot index were all recorded after application of PGPR strains (fluorescent *Pseudomonas* spp. RB50 and RB31).

» Also beneficial is the application of 2 kilogram of *Pseudomonas fluorescens* combined with 300 kg of compost.

References

Thangavel, K., Baskaran, R. M., Rajavel, D. S., Manisegaran, S., & Suresh, K. (2012). Field evaluation of biointensive module for the management of major insect pests of coleus. Journal of Biopesticides, 5, 159.

Singh, R., Gangwar, S. P., Singh, D., Singh, R., Pandey, R., & Kalra, A. (2011). Medicinal plant Coleus forskohlii Briq.: disease and management. Medicinal Plants-International Journal of Phytomedicines and Related Industries, 3(1), 1-7.

17 Isabgol

INSECT PESTS

Aphid, *Aphis gossypii*

Damage symptoms

- Aphids, both green and brown in appearance, quickly colonise plants, especially young shoots and the undersides of leaves.
- Adult and juvenile stages of this pest feed on the plant's sap, causing the affected tissues to curl, wither, and develop more slowly than they otherwise would.
- Curled, withered leaves fall off when they are attacked.
- Infestation is so bad that the crop has to be destroyed.

Pest identification

- Larvae range in colour from grey to green; they typically have dark markings on the head, thorax, and wing pads and a green belly.
- All adults lack wings and range in colour from light to dark green with black markings on the head and chest.

Management

- The aphid population can be efficiently reduced by spraying 1% Neem or Pongamia oil.
- Among Neem products, the benefit cost ratio for the 2% crude extract treatment was the next highest.

» Aphids could be controlled using a 0.05% concentration of dimethoate, oxademeton methyl, or fenthion.

» With a benefit-to-cost ratio of 0.9, Oxydemeton-Methyl 0.025% proved to be the most effective treatment for pest management on Isabgol cv. Jawahar (16.9).

» When given enough time, Coccinellids and Syrphids can effectively control aphid populations. *Coccinella septempunctata, C. transversalis, Cheilomenes sexmaculata (= Menochilus sexmaculatus), Scymnus* sp., *S. castaneus,* and *S. latemaculatus (*all coccinellids*), Paragus yerburiensis,* and *Sphaerophoria macrogaster* (both syrphids), *Chrysopa* (Tandon, 1994).

» *Aphelinus gossypii*, a parasitoid, reached its highest prevalence at this time of maximum population.

DISEASES

Downy mildew, *Peronospora plantaginis, P. alta*

Symptoms

» Leaves become chlorotic on the upper surface and mycelial growth that resembles frosty ash appears on the lower surface.

» A few stages later, the infected leaves turn a chlorotic brown and curl, wilt, and dry out.

» chlorotic streaks appear, running parallel to the leaf midrib.

» By the end, the entire leaf has turned necrotic, and in extreme cases, the entire plant has a blighted aspect.

» The damaged leaves get thicker as a result of the high oospore production.

Favorable conditions

» Between 2:00 and 8:00 in the morning and again at 10:00 at night, when the temperature is approximately 20 °C and the relative humidity is around 89%, sporangial germination occurs most successfully.

Survival and spread

» This disease spreads from soil to seeds.

Management

» The maximum seed rate for a crop planted in November is 2.5 kg/ha.

- Plants that are planted in December have a better chance of avoiding the disease.
- Seeds were soaked in 0.25 g/10 litres of water of Aureofungin the night before planting, and three sprays of 15 g/ha of Aureofungin were used to effectively manage the disease.
- Treatment of seeds with 5 g/kg of Apron SD, followed by 3 sprays with either 0.2% Captafol, 0.05% Metalaxyl, or 0.3% Mancozeb, was sufficient to keep the disease under control.
- Treatment of seed with 0.3% Mancozeb or Metalaxyl at 4 g/kg seed, followed by three 15-day interval sprays of 0.2% Ridomil or Metalaxyl at 2 g/L water beginning at the first sign of disease.
- At 30 DAS, spray Mancozeb (Dithane M-45) or Zineb (Dithane Z-78) at 2.0 to
- 2.5 g/L or Bordeaux mixture (6-3-50) at 600 to 800 L/ha, and repeat the process twice more, with a 14- to 16-day gap between applications.
- Downy mildew resistance was reported in five cvs (MIB-123, MIB-124, AMB-2, P-6, and P-80).
- During the early stages of a pest assault, should remove the affected plant portions, spray with 0.05% Dimethoate (30 EC), 0.05% Methyl Demeton (25 EC), or 0.1% Malathion (50 EC), and release coccinellid predators.

Powdery mildew, *Erysiphe cichoracearum, Podosphaera leucotricha*

Symptoms

- Powdery mildew manifests as a white, powdery growth on the upper surface of leaves, typically around the time of flowering.
- Diseased plants initially show a whitish or greyish speck on their leaves, which grows larger and eventually covers the entire plant in a powdery substance reminiscent of talcum powder.

Management

- When the first signs of disease develop, spraying with 0.2% Karathane or Wettable Sulfur (such as 'Karathane' and 'Sulfex') two or three times at 15-day intervals is effective in eradicating the ailment.

References

Manikandan, R., Singh, A. K., & Singh, H. M. (2007). Studies on Bionomics of Aphis gossypii infesting Plantago ovata. Annals of Plant Protection Sciences, 15(1), 243-244.

Varghese, T. S., Tripathy, V., & Kumar, J. (2018). Dissipation of imidacloprid on white psyllium (Plantago ovata) under semi arid conditions. Proceedings of the National Academy of Sciences, India Section B: Biological Sciences, 88, 547-551.

Khinchi, S. K., Kumawat, K. C., & Bagdi, D. L. (2021). Biotic parameters of ladybird beetle, Coccinella septempunctata Linn. On aphid, Rhopalosiphum maidis (Fitch) of blond psyllium, Plantago ovata Forsk.

Mandal, K., Saravanan, R., Maiti, S., & Kothari, I. L. (2009). Effect of downy mildew disease on photosynthesis and chlorophyll fluorescence in Plantago ovata Forsk. Journal of Plant Diseases and Protection, 116(4), 164.

Samad, A., Ajayakumar, P. V., Zaim, M., Lal, R. K., & Khaliq, A. (2002). Phytoplasma associated with little leaf disease of psyllium (Plantago ovata) in India. Journal of herbs, spices & medicinal plants, 9(1), 55-61.

18

Long Pepper

INSECT PESTS

Mealy bugs, *Dysmicoccus* sp.

Damage symptoms

- » Feeding on a plant's sap, mealybugs infest the plant's roots.
- » The disease is causing yellowing of the affected plant.
- » The plant's development has stalled.

Favorable conditions

- » Summer is peak season for extreme weather.

Management

- » Neem cake at a rate of 150 kg per hectare applied to the soil.
- » Dimethoate 2 ml/L soil drench around the base of the damaged plants.
- » Chlorpyriphos applied to the soil at a concentration of 3 ml/L around the plant's root zone.

Tea mosquito bug, *Helopeltis theivora*

Damage symptoms

- » Both nymphs and adults feed voraciously on the plant, causing serious damage to the young leaves.
- » After a while, the leaf lamina will acquire necrotic lesions around the feeding puncture, looking like a splattering of bullet holes.

Pest identification

- » The adult has a black body, red thorax, black and white abdomen, and greenish brown wings.

Management

- » Evening applications of 2 ml/L of Endosulfan, Phosalone, or Monocrotophos in water provide the best control.
- » For successful control of mealy bugs and *Helopeltis theivora,* damage caused by these pests to young, vulnerable plants can be mitigated by spraying the plants with 4% Neem seed kernel extract or with another pesticide based on neem.

DISEASES

Anthracnose, *Colletotrichum gloeosporioides*

Symptoms

- » Induce leaf decay during the monsoon season.
- » To bring about necrotic lesions on leaves during the summer.
- » The fungus causes decay in plant tissues, such as leaves and vines.

Management

- » To lessen the damage this disease does, spray a Bordeaux mixture of 0.5% every two weeks and saturate the soil with a 1% solution once a month.
- » The disease can also be efficiently controlled by spraying three times with 0.2% Difolatan every 15 days.

Stem rot, *Sclerotium rolfsii*

Symptoms

- » The months of June through September are the wettest. Stems develop a white, mushy rot of decaying tissue.

Management

- » Spraying a Bordeaux mixture of 0.5% every two weeks and saturating the soil with a 1% solution once a month.
- » Copper oxychloride and bacitracin, both at 0.2%, also work well.

References

Hoza, G., Dobrin, I., Dinu, M., Becherescu, A., Ilie, V., & Bezdadea, I. C. (2016). Research regarding the use of natural predators for the control of pests for pepper in protected culture. Agriculture and Agricultural Science Procedia, 10, 192-197.

Bhat, A. I., Biju, C. N., Srinivasan, V., Ankegowda, S. J., & Krishnamurthy, K. S. (2018). Current status of viral diseases affecting black pepper and cardamom.

Parisi, M., Alioto, D., & Tripodi, P. (2020). Overview of biotic stresses in pepper (Capsicum spp.): Sources of genetic resistance, molecular breeding and genomics. International Journal of Molecular Sciences, 21(7), 2587.

19 Medicinal Yam

INSECT PESTS

Aphids, *Aphis gossypii*

Damage symptoms

- Aphids are a pest that can ruin seedlings.
- Sap from young vine leaves and stems is sucked by nymphs and adults.
- This has a chilling effect on future expansion.

Pest identification

- Small, round, and anywhere from a muddy brown to a pitch black.

Management

- Use 5% NSKE or 0.5% Neem oil in a spray bottle.
- Protect predators and parasites by preserving their natural habitats.
- Use 1.5 ml of Monocrotophos, 3 g of Wettable Sulfur, and 0.5 ml of Teepol per 1 litre of water in a spray bottle.

Red spider mite, *Tetranychus* sp.

Damage symptoms

- The petiole that attaches to the base of the luminaire is infested with red spider mites.
- Necrotic lesions appear on the upper laminar surface of leaves.
- Both *D. floribunda* and *D. composita* are susceptible to fungal infection after suffering feeding injuries.

Pest identification

» All life stages, from nymphs to adults, are bright red.

» Round, white eggs are placed on the underside of the leaves.

Management

» This pest can be eliminated with a spray mixture of 3 gm of Wettable Sulfur and 0.5 ml of Teepol per litre of water.

DISEASES

Anthracnose, *Colletotrichum gloeosporoides*

Symptoms

» Leaf necrosis; stem dieback; withered leaves with a burned look; small, dark brown spots or black lesions encircled by a chlorotic halo.

» As a result of the blackening and death of the yam's leaves and young shoots, the crop is stunted, undersized, and misshapen.

Survival

» Viruses can survive the winter in dead plants.

Management

» Superior field sanitation.

» In order to stop the spread of the disease, it is imperative that promptly remove any diseased plants and plant debris from the field.

» A spray of Benomyl or Maneb will stop the spread of anthracnose.

References

Akoroda, M. O. (1993). Yams: Dioscorea spp. In Genetic improvement of vegetable crops (pp. 717-733). Pergamon.

Amusa, N. A., Adigbite, A. A., Muhammed, S., & Baiyewu, R. A. (2003). Yam diseases and its management in Nigeria. African Journal of Biotechnology, 2(12), 497-502.

Atiri, G. I., Winter, S., & Alabi, O. J. (2003). Yam. In virus and virus-like diseases of Major Crops in Developing Countries (pp. 249-268). Dordrecht: Springer Netherlands.

Reddy, P. P., & Reddy, P. P. (2015). Yams, Dioscorea spp. Plant Protection in Tropical Root and Tuber Crops, 193-233.

Jayaprakas, C. A., & Harish, E. R. (2022). Pests and Their Management in Minor Tuber Crops: (Chinese Potato, Yams, Taro, Elephant Foot Yam and Tannia). Trends in Horticultural Entomology, 1109-1137.

20 Opium Poppy

INSECT PESTS

Cutworm, *Agrotis sulfusa*

Damage symptoms

- Larvae feed on the roots and shoots of mature plants, killing them at ground level, at night.
- Damage causes growth inhibition and decreased harvests.

Pest identification

- Larvae are a dark brown tint with a reddish head.
- The bodies of adult moths are thick, fatty, and greenish brown or dark. It has a dark forewing with white patches and a pale rear wing.

Management

- Pupae are vulnerable throughout the warm months and can be uncovered by deep ploughing.
- Submerging the infected area with water.
- Eliminating the larvae by hand.
- In the evening, the larvae in the field crawl behind the thick mats of green grass for protection. The next morning, remove the pile and all the larvae in it.
- It's effective to spray 5% NSKE.
- Before planting, dust with Malathion 50 EC.
- Chlorpyriphos 20 EC diluted in water at a rate of 2 ml/L is sprayed.

- Larvae can be eradicated from a field by scattering a poison bait consisting of rice bran, jaggery, and Carbaryl/Chlorpyriphos (10:1:0.5 ratio).
- Both Methyl Parathion and Phenthoate, when sprayed at a concentration of 0.05%, are efficient pesticides.

Capsule borer, *Helicoverpa armigera*

Damage symptoms

- Significant damage is done to poppy fields by this insect.
- Round, creamy eggs are placed on young leaves, and the hatchlings will subsequently bore holes into the developing spikes and capsules.
- Its diet consists primarily of flower pods and seeds.

Pest identification

- Larvae have a bluish-green hue and dark brown or grey stripes running down their bodies.
- The adult moth is about the size of a dime and has brown wings with a black 'V' shaped speck and a dull black border.

Management

- Larvae are selected by hand.
- 12 pheromone traps per hectare (ha) used.
- Apply a spray containing either 5% NSKE or 1% Neem oil to prevent the larvae from feeding.
- Spray with 0.01% Fenvalerate, 0.05% Methyl Parathion, 0.05% Quinalphos, or 0.07% Endosulfan.
- The nuclear polyhedrosis virus of *H. armigera* should be sprayed at a rate of 250 LE/ha during the larval stage.
- Apply a spray formulation of Bt at the rate of 2 g (2 ml) per litre of water.
- At least two applications of 5 ml/ha of *Trichogramma chilonis* spaced 15 days apart can be made throughout the egg-laying cycle.
- *Banchopsis ruficornis*, *Bracon brevicornis*, *Campoletis chlorideae*, *Carcelia illota*, *Enicospilus* sp., *Eriborus* sp., *Goniophthalmus halli*, *Trichogramma chilonis*, and *Trichogrammatoidea armigera* are all examples of common parasitoids.
- Animals such as the *Delta* sp., *Chrysoperla carnea*, *Orius* sp., *Menochilus*

sexmaculata, *Rhynocoris* sp., *Geocoris* sp., and *Nabis* sp.

» The populations of *H. armigera* are controlled in part by HaNPV, *Bacillus thuringiensis, Beauveria bassiana*, and *Metarrhizium anisopliae.*

Aphid, *Myzus persicae*

Damage symptoms

» Adults and nymphs alike feed on the sap of leaves, causing damage to developing stems and branches.

» Deformed, withered, and stunted growth are the results for affected shoots.

» The *M. persicae* population was found to be largest on leaves, and lowest on capsules.

Pest identification

» Yellowish green as an adult, occasionally red.

Management

» Insect infestation was lower in the crop that was planted in early October.

» Aphids can be efficiently controlled by spraying the plants twice with NSKE 5% or 2% Neem/Pongamia oil at intervals of 10-14 days.

» Aphid populations can also be drastically cut with sprays of 0.05% Dimethoate, 0.05% Phenthoate, or 0.05% Profenphos.

» Promote the presence of natural adversaries such as the green lacewing (*Chrysoperla carnea*).

DISEASES

Downy mildew, *Peronospora arborescens*

Symptoms

» When a plant has a systemic infection, it shows two distinct sets of symptoms: the lower part is healthy, while the top part is sick.

» The upper portion of the plant remains infected, and the leaves remain tiny, chlorotic, and curl downward at the edges.

» All over the underside of the sick leaves, sporulation is very dense.

» In most cases, buds fail to fully mature and open.

- After an infection, if a capsule form at all, it will eventually dry out, shrink, and develop wrinkles.
- There is abundant, transient sporulation on the peduncle, the buds, and the capsules.
- In place of mature seeds, rudimentary brownish-black structures form.
- These pills do not contain as much opium as others.
- Infected plants tend to die off quickly.

Management

- Early seeded crop had the least downy mildew incidence in Udaipur.
- Disease incidence and severity diminished as plant density increased.
- Metalaxyl sprays at 0.2% applied 35, 55, 75, and 95 days after planting were found to be the most effective in lowering both primary and secondary disease incidence.
- Sprays of Ridomil MZ, Cuman-L, and 0.2% Mancozeb, along with apricot seed treatment (8-10 g/ha), proved to be particularly successful in preventing the spread of the disease.

References

Kubelková, D., & Špak, J. (1999). Virus diseases of poppy (Papaver somniferum L.) and some other species of the Papaveraceae family-a review. Plant Protection Science, 35(1), 33-36.

Kishore, R. A. J., Tripathi, R. D., Johri, J. K., & Shukla, D. S. (1985). Some new fungal diseases of opium poppy (Papaver somniferum L.). Indian Journal of Plant Pathology, 3(2), 213-217.

Miciński, B., & Przybylska, M. (1983). Preliminary experiments on virus diseases of poppy (Papaver somniferum L.) in Poland. Zeszyty Problemowe Postępów Nauk Rolniczych, (291), 221-225.

Giray, H. (1985). A list of pests of popy (Papaver somniferum L.) in Turkey, with notes on the types of damage of the important species. Turkish Journal of Entomology, 9(2).

Türkoğlu, T., & Fidan, Ü. (2008). Preliminary studies on the establishment of the pest and diseases of opium poppy (Papaver somniferum L.) in Aegean region. Plant Protection Bulletin.

21 Periwinkle

INSECT PESTS

Aphids

Damage symptoms

- When plants are fed, their growth is stunted.
- Turning yellow or brown is possible for periwinkle leaves.
- Under strong light, they wilt and may even pucker and curl.
- Aphids eat the leaves, but their size prevents them from leaving noticeable holes.
- Curled, twisted, or yellowed leaves.
- Aphids can transmit diseases to plants, causing harm to the leaves and even stunted growth, and they can leave behind a sticky residue that can look like black soot.

Management

- Knocking the aphids off the vinca vines with a forceful stream of water might be sufficient for eradicating a light infestation.
- Every three to five days, spray with insecticide soap.
- Neem products should be sprayed on aphids and foliage surfaces twice or three times every 10 days.
- Aphids are prey for lady beetles and other beneficial insects.
- Parasitic wasps, like the *Lysiphlebus testaceipes*, are responsible for the extinction of aphids.

Grasshopper, *Acrida exalta*

Damage symptoms

» Both juveniles and adults are destructive to plant life, eating holes in leaves and snipping young shoots.

Management

» Protect plants from caterpillars with a spray of Dichlorvos 0.05% or Carbaryl 50 WP 0.1%.

Sphingid larvae, *Dilephila nerii*

Damage symptoms

» Larvae do significant harm to plants because they consume so many of the plant's leaves, buds, and blossoms.

Management

» Pupae are vulnerable to avian predation because the soil is dug up to expose them.

» Caterpillars are hand-picked and wiped off.

» One light trap per hectare is installed.

» A trap plant, Nerium (*Nerium olender*), is planted around the field.

» Collect larvae and kill them all.

» Apply a spray containing 2 ml of Methyl parathion or Chlorpyriphos 20 EC per litre of water.

Leaf eating caterpillar, *Anomis flava*

Damage symptoms

» Intense leaf loss occurs as a result of larval feeding.

Pest identification

» As they move, larvae crawl in a cyclical pattern.

» The adult has a stocky build and dark brown patches on its brown wings.

Management

» After harvesting, plough the field to eliminate any remaining pupae.

- » Green growing practises.
- » Restoration work is the cutting away and burning of diseased or dying plants.
- » Control measures include removing and killing caterpillars and their egg masses by hand.
- » For efficient management of caterpillar damage, spray a solution of Dichlorvos 0.05% or Carbaryl 50 WP at a concentration of 2-3 g/L of water.
- » Letting loose parasitoids like *Tricholyga sorbillane* and *Litomastrix gopimohani.*

DISEASES

Stem blight/root rot, *Phytophthora nicotianae*

Symptoms

- » Yellowing and burning of leaves, slow growth and stunting, withering and dying are all signs of root rot.
- » Spots on leaves are water-soaked, uneven regions that first appear on the underside of the leaf and then spread to destroy the entire leaf.
- » The rot spreads from the leaves to the stem, eventually killing off whole portions or branches.
- » Over time, the fungus feeds and destroys the root system, infecting and killing off small feeder roots first and eventually progressing to larger roots.
- » The plants infected with the disease may be stunted, weak, or appear to be malnourished.
- » Root rot causes plants to have diminished root systems and causes individual roots to shed their outer tissue, exposing the inner core.

Survival and spread

- » To avoid dying off without a host, chlamydospores develop thicker walls, allowing them to stay dormant in soil for months.
- » Zoospores are splattered from sporangia to healthy leaves or other plants.
- » Contaminated soil and sick plants can travel great distances.

Management

- » Sow only wholesome seed and keep an eye out for signs of disease in your seedlings.

- The primary preventative approach is water management (water only as needed).
- Also crucial is the provision of suitable drainage.
- Increasing airflow will help dry out the leaves.
- Increasing drainage is possible when organic materials, like composted pine bark, are added to the soil.
- Remove and burn the diseased vegetation.
- The use of the contact fungicide mancozeb has been shown to minimise the incidence of leaf spots.
- Leaves and young stems can be safeguarded by using Metalaxyl, a systemic fungicide.

Stem and root rot, *Rhizoctonia solani*

Symptoms

- Damping off is a common fungal disease that affects young plants, but it can also affect the roots of mature plants.
- Plants may show symptoms such as withering, yellowing, and stunting.
- Symptoms include stunted growth, brown lesions on the roots, yellowing leaves, and wilting even when soil moisture levels are adequate.
- Infected plants may wilt quickly in the heat of the summer, and even more watering may not save them.

Management

- Make sure the plants are healthy and green but not wilted.
- Water supplementary only as required.
- Change to a hardier plant every so often, like marigold.
- In the case of root diseases, it is essential to completely uproot infected plants and kill them.
- Applying a soil and root drench with Banrot and Abound before the first signs of disease develop is an effective method of disease prevention.
- Apply fungicides like iprodione (Rovral or), chlorothalonil (Daconil), and thiophanate methyl (Mefenoxam).

Gray mold (Botrytis blight), *Botrytis cinerea*

Symptoms

- Leaves, stems, flowers, flower buds, seeds, seedlings, and bulbs are all susceptible to Botrytis.
- It is common for the fungus to begin colonising dead plant components before moving on to healthy ones.
- The diseased portion of a plant will be very dark or blighted and will be covered in a dense layer of grey fuzz.
- Leaf spots and blights, stem cankers, stem rots, and damping off of seedlings are just some of the rapid-progressing symptoms.
- *Phytophthora* spp. also cause cankers on stems that look very similar.
- On decaying plant matter, a dense grey fungal growth may develop, punctuated by tiny black flecks (sclerotia).

Favorable conditions

- Springtime rain and temperatures between 13 and 18 °C encourage the growth of disease, which is especially common in overly fertilised landscaping.

Survival and spread

- The fungus' sclerotia serve as a winter home for the sickness.
- Infected plants serve as a wintertime refuge for the fungus.
- In wet weather, the mycelia may release clouds of airborne spores, which can further spread the disease.

Management

- Do not leave seriously infected plants in the garden.
- One should only water when absolutely necessary. If you must water, do so early in the morning so that the leaves have time to dry before nightfall and avoid watering from above.
- Careful plant spacing will allow for increased air flow within the plant canopy.
- Products containing iprodione (Rovral), potassium bicarbonate, mancozeb, and thiophanate methyl are all effective fungicides.

Stem blight, *Phoma exigua*

Symptoms

- Runners are frequently girdled and killed when black lesions appear initially on the leaves and stems.
- The black tissue contains fruiting bodies of the fungus hardly bigger than pinheads.
- The trunk and lower branches of the plant become black and die.
- This fungus causes the stems and leaves to become brown and eventually die.

Favorable conditions

- This mould grows within persistently damp conditions.

Survival

- The fungus develops rapidly in damp environments, such as dense plantings of ground coverings, particularly in the spring when there is abundant moisture and succulent tissue.

Management

- It's best to not water the lawn with a sprinkler gun.
- Touch plants when they are completely dry. This aids in reducing the likelihood of further disease transmission.
- Remove all the contaminated plants, including any old runners that might be hiding in the background. Sweep the bed of dead leaves to improve air circulation.
- Spray a preventative fungicide containing copper on young growth in the spring to keep it healthy and disease-free.
- A drenching spray of a fungicide (Iprodione or Propiconazole) should be used.

Rust, *Puccinia vincae*

Symptoms

- Leaf spots develop on the upper surface of the leaf and the leaves turn chlorotic.
- Dark uredinia can be found on the undersides of leaves, which is also where telia are developed.

Management

- Eliminate and dispose of any contaminated plants or plant pieces.
- Not water plants directly overhead or keep them soggy for long periods of time.
- Use Zyban WSB at 24 oz/100 gal water, Terraguard SC at 4 to 8 fl oz/100 gal water, Pageant at 12 to 18 oz/100 gal water, etc.

Tomato Spotted Wilt Virus (TSWV)

Symptoms

- Yellowing, stunting, and distorted growth are only some of the symptoms.
- Thrips are the vectors of transmission for TSWV.

Management

- Before buying plants, make sure there are no signs of infection.
- Diseased plants cannot be treated and contribute to the spread of the disease, thus they must be removed from flower beds.
- Stay away from planting vinca near vegetable gardens because many vegetables are also susceptible.
- Because many weeds are capable of harbouring both thrips and the virus, weed control is also crucial.

Little leaf (Phytoplasma)

Symptoms

- Mycoplasma infections can cause tiny leaf disease, which causes some plants to grow slowly and develop rosettes on their leaves.
- Insects serve as "vectors" for the disease's propagation (leaf hopper, *Orosius albicinctus*).

Management

- Infected plants can be destroyed to stop the disease from spreading.
- At 15-day intervals when the infection is widespread, spraying with Metacid or Nuvacron can help reduce the spread of the disease and its vector.

References

Davis, R. E., Harrison, N. A., Zhao, Y., Wei, W., & Dally, E. L. (2016). 'Candidatus Phytoplasma hispanicum', a novel taxon associated with Mexican periwinkle virescence disease of Catharanthus roseus. International Journal of Systematic and Evolutionary Microbiology, 66(9), 3463-3467.

Kulkarni, R. N., Kalra, A., & Ravindra, N. S. (1992). Integration of soil solarization with host resistance in the control of dieback and collar and root rot diseases of periwinkle. Tropical agriculture, 69(3), 217-222.

Kulkarni, R. N., & Ravindra, N. S. (1997). Integration of host resistance with fungicide in the control of dieback disease of periwinkle. Tropical agriculture.

Chen, W. Y., Huang, Y. C., Tsai, M. L., & Lin, C. P. (2011). Detection and identification of a new phytoplasma associated with periwinkle leaf yellowing disease in Taiwan. Australasian Plant Pathology, 40, 476-483.

Basavand, E., Firouzianbandpey, S., & Rahimian, H. (2022). Periwinkle leaf spots and stem lesions caused by Xanthomonas campestris. European Journal of Plant Pathology, 164(2), 167-176.

Mishra, P. P., & Sasmal, A. (2020). Pests of Medicinal and Aromatic Plants and Their Integrated Pest Management Techniques. Applied Entomology and Zoology, 15.

MH, S. K., & Srinivas, M. P. (2018). Pests attacking medicinal and aromatic plants in India: a review. Journal of Entomology and Zoology Studie, 6(5), 201-205.

Tripathi, A. K. (2018). Pests of Medicinal and Aromatic Plants. Pests and Their Management, 719-822.

22 Sarpagandha

INSECT PESTS

Cutworm, *Agrotis* sp.

Damage symptoms

- The young seedlings are prey to larva.
- Caterpillars typically come out at night and hide in the shadows near plants.
- For a few days, tiny caterpillars munch en masse on the plant's leaves before tunnelling below.
- Eats young seedlings by severing them just above the soil.
- Lower lying, more water-logged places are hit the hardest.

Pest identification

- Eggs are round, ribbed, and creamy white in colour.

Management

- During the summer, fork the ground to expose larvae and pupae to avian predators.
- To lure adult moths into light traps during summer, use the appropriate seasonal language.
- Pheromone traps should be set up at a rate of 20 per hectare to track male moths and catch them for study.
- Daytime irrigation of the land will attract birds that will feast on the exposed larva.

- Spray 2 ml/L of Chlorpyriphos or Endosulfan into the soil around the plant's collar area once daily for seven days after planting.

Epilachna beetle, *Henosepilachna vigintioctopunctata*

Damage symptoms

- While feeding, grubs and adults scrape chlorophyll from the epidermal layers of leaves, leaving behind a skeletonized form that eventually dries out and falls off.

Pest identification

- Yellow in hue, grubs are stocky and covered in spines.
- The adults of this species are spherical in shape, and their bluish-brown colouring is speckled with black dots.

Management

- Collect the grubs and kill the adults.
- Apply a mixture of Carbaryl 50 WP 3 g/L, Wettable Sulfur 2 g/L, and Endosulfan 35 EC 2 ml/L.

Sphingid caterpillar, *Deilephila nerii*

Damage symptoms

- Caterpillars produce defoliation by munching on leaves.

Pest identification

- The larva has a spine in the anal area and is somewhat stocky overall.
- The adult moth has a strong green colour and is marked with yellow.

Management

- Puppy death from bird predation caused by digging holes in the soil.
- Caterpillars are hand-picked and wiped off.
- A 1/ha installation rate for light traps.
- *Nerium olender* plants are planted as a trap plant around the perimeter of the field.

Ash weevil, *Myllocerus viridanus*

Damage symptoms

- The adults use their sharp beaks to eat the leaves' edges, which results in a notch being cut into the leaves.
- Weevils, or grubs, eat plant roots, leading to plant death.

Management

- Adults will be gathered and subsequently annihilated.
- Treatment with a spray containing either 0.2% carbaryl or 0.1% malathion.
- Dust incorporation or soil drenching with 0.1% Chlorpyriphos has been shown to be effective in reducing soil-borne pathogens.

Mealy bug, *Coccidohystrix insolita*

Damage symptoms

- Adults and juveniles alike can be found feeding on the sap from the undersides of leaves.
- Infected plants wilt and turn a sickly yellow colour.
- A black mould appears as a result of honeydew's excretion.

Pest identification

- Little, round, and squishy, mealy bugs have white mealy wax all over their bodies.

Management

- Elimination of infected organs at an early stage of infection.
- Spraying with either 0.05% dimethoate, 0.05% methyl demeton EC, or 0.1% malathion EC.
- Dispersal of predatory Coccinellids into the wild.
- Spraying 5 ml of Neem oil or 25 g of Fish oil rosin soap per litre of water on the crop at its latter stages is recommended.

Scales, *Saissetia coffeae*

Damage symptoms

- The undersides of the leaves, including the midribs and midveins, are infested with brownish-black scales.

- Petioles and young shoots are protected, too.
- The leaves become yellow and fall off because their sap has been drained.
- Plant growth is stunted.

Pest identification

- There is a yellow H-shaped mark on the bodies of female nymphs.
- The mature insect is initially green but eventually turns a dark brown. It is only 2 mm in length and has a robust convex helmet.

Management

- Infested plant portions are cut off and burned.
- Apply a 0.05% solution of Methyl Parathion 50 EC or Ethion 50 EC every 15 days.

Chafer beetle, *Anomala polita*

Damage symptoms

- The initial rains usually do the most damage.
- Seedlings are killed when cockchafer grubs feed on them around 2 centimetres below the hypocotyls.
- At night, adults eat many types of plant life, including buds and blossoms.

Pest identification

- The grub's body is white, but its head is a dark brown.
- An adult has a small to medium body that is a metallic green colour with a pygedium that is visible.

Management

- In the evening, spray 1 ml of Methyl Parathion 50 EC or 2 ml of Quinalphos 25 EC per L of water.
- Mixing Phorate granules into the soil during nursery preparation can help prevent grub attacks.

DISEASES

Blossom blight, *Colletotrichum capsici*

Symptoms

- Resulting in a significant drop in seed quality.
- Disease has a negative impact on the sarpagandha plantation because it causes severe infection on inflorescence, which ultimately kills the affected flowers before they fully bloom.
- Decapitation and the inability to set seeds are other outcomes of the bloom blight.

Management

- Treatment with sprays of Mancozeb at 0.3% and Carbendazim at 0.15% resulted in the strongest protection against blossom blight disease (> 80%) in both years, as well as the healthiest plants and the highest seed yield per plant.
- A spray concentration of 0.25 percent Mancozeb or 0.25 percent Carbendazim offered 87.8 and 84.6 percent protection, respectively.

References

Shukla, R. S., Alam, M., Sattar, A., Abdul-Khaliq, & Singh, H. N. (2006). First report of Rhizopus stolonifer causing inflorescence and fruit rot of Rauvolfia serpentina in India. EPPO Bulletin, 36(1), 11-13.

Mehrotra, M. D., & Thapar, H. S. (1990). Rhizoctonia leaf spotting and blight of Rauvolfia serpentina, a new disease from India. Indian Forester, 116(5), 372-374.

Dadwal, V. S., Bhartiya, S., & Verma, R. K. (2010). A New Top Dying Disease of Rauvolfia serpentina, Caused by Lasiodiplodia theobromae. J Mycol Plant Pathol, 41(2), 246-248.

Dadwal, V. S., & Nisha, S. (2013). Diseases of Rauvolfia serpentina and their biocontrol. Indian Journal of Forestry, 36(2), 197-204.

Khan, O. A., Padhi, J., & Padhy, D. (2020). Beneficial and harmful fauna associated with Sarpagandha (Rauvolfia serpentina) in Odisha. Journal of Pharmacognosy and Phytochemistry, 9(4S), 20-21.

Suganthy, M., Baskaran, R. M., & Mani, M. (2022). Pests and Their Management in Medicinal Plants: (Muskdana, Aonla, Ashwagandha, Black Datura, Black

Nightshade, Coleus forskohlii, Gymnema sylvestre, Isabgol, Glory Lily, Long Pepper, Noni, Opium Poppy, Periwinkle, Phyllanthus amarus and Phyllanthus niruri, Psoralea, Sarpagandha, Senna, Madhuca longifolia, Mimusops elengi, Wrightia tinctoria, etc.). Trends in Horticultural Entomology, 1255-1303.

23 Senna

INSECT PESTS

Green leaf eating caterpillar, *Catopsilia pyranthae, Eurema hacabe*

Damage symptoms

- Damage and defoliation are severe because the larvae eat on tender young leaves.

Pest identification

- Larvae are cylinder-shaped and green with yellow stripes.
- In adults, the upper side of the wings are white.
- A male can be identified by the presence of a black tip on the leading edge of his wing.
- The female has a brown speckling and a black margin around each wing.

Management

- Handpicking.
- Pods that have been invaded must be destroyed along with the larvae inside.
- Use a spray bottle filled with water and either a 5% NSKE or a 2% Bt formulation to kill the pests.
- Using a spray containing either 0.5% carbaryl or 0.05% endosulfan.
- Applying a 3% Neem oil spray in the beginning stages.
- For example, whereas *Trichogramma chilonis* was responsible for parasitizing the vast majority of *C. pyranthe* eggs, *Charops obtusus* and *Apanteles* sp. were

responsible for parasitizing the vast majority of *C. pyranthe* larvae, with maximum rates of 3125% and 2957%, respectively.

» In addition to chrysopids like *Brinkochrysa seclestes* and chrysopoids like *Acridotheres tristis*, insectivorous birds like Passer *domesticus indicus* and *Acridotheres tristis* also feed on larvae.

Pod borer, *Etiella zinckenella*

Damage symptoms

» The pod borer punctures the pods to get at the seed inside.

Pest identification

» The tiny, white, elliptical egg is placed at the point where the calyx meets the pod.

» The larvae are rosy-pink on top and a drab green underneath.

» The wings of an adult greyish brown moth have a thin white stripe.

Management

» Handpicking.

» Pods that have been invaded must be destroyed along with the larvae inside.

» In the middle of the flowering phase, spray Bt at a rate of 0.1%.

» *Trichogramma chilonis* at 5 ml per hectare released in the field.

» Make use of a 1/ha light trap.

» Using a spray containing either 0.5% carbaryl or 0.05% endosulfan.

DISEASES

Leaf-spot, *Alternaria alternata*

Symptoms

» There are small grey dots all throughout the leaves.

» The spots grow in size until they become complete circles.

» If the infestation is bad enough, the leaves will fall off the plant, which will result in a low harvest.

» Sennoside levels in leaves are inversely related to disease severity.

Favorable conditions

- Conditions conducive to the spread of the disease include cloudy, humid weather.

Management

- A 0.25 percent concentration of nimin in a spray reduced the severity of the sickness to a minimum.
- Treatment with a spray containing 0.15 percent Mancozeb (Dithane M-45) was performed three times over a period of five to six weeks at two-week intervals.
- Bordeaux combination, Benomyl, Captafol, and Copper oxychloride are all suggested as sprays to use.

Rust, *Ravenelia* sp.

Symptoms

- Bring about the withering of plants and trees, death of leaves and branches.

Management

- Damaged plant portions can be pruned.
- Use a spray of Bayleton, Banner, Heritage, or Pageant.

References

Baskaran, R. K., Senthilkumaran, S., Rajavel, D. S., Shanthi, M., & Suresh, K. (2008). Yield loss by Major Insect Pests in Senna (Senna angustifolia). Annals of Plant Protection Sciences, 16(2), 495-496.

Veeraragavan, S., Duraisamy, R., & Mani, S. (2018). Prevalence and seasonality of insect pests in medicinally important plant Senna alata L. under tropical climate in the Coromandel Coast of India. Geology, Ecology, and Landscapes, 2(3), 177-187.

Marimuthu, T., Suganthy, M., & Nakkeeran, S. (2018). Common pests and diseases of medicinal plants and strategies to manage them. New Age Herbals: Resource, Quality and Pharmacognosy, 289-312.

Baskaran, R. K., Senthilkumaran, S., Rajavel, D. S., Shanthi, M., & Suresh, K. (2009). Effect of Organic Sources of Nutrients on Management of Sucking Pests of Cassia angustifolia Vahl. Annals of Plant Protection Sciences, 17(1), 32-36.

24 Soda Apple

INSECT PESTS

Leaf beetle, *Gratiana boliviana*

Damage symptoms

- Adult and larval beetles consume tropical soda apple leaves, chew holes in them, and defoliate plants to the point that photosynthesis is stunted,leaving them vulnerableto infections and even extinction.
- Both the larvae and the adults feed on the plant's leaves, weakening the plant and slowing its growth and its ability to produce fruit.
- Defoliation from beetles can range from 30 to 100 percent, drastically reducing a plant's ability to bear fruit (from 40 to 60 per plant to none or very few).

Pest identification

- The eggs are brown and between 1 and 1.8 mm in length, and they come in a papery casing.
- Larvae are a spiky, pale green colour.The cast skins and frass used as disguise bylater instars are easily recognisable.
- Pupae are between 5 and 6 mm in length,pale green in colour,spiky,flattened, and motionless. As their name implies, pupae are typically located on the underside ofleaves.
- The length and width of an adult are roughly 6 and 4 to 5 millimetres, respectively. Adults in their reproductive stage are a dark green, while those in their diapause (winter) stage are a lighter brown.

Management

- *Coleosoma acutiventer, Peucetia viridans, Misumenops celer, Acanthepeira stellate, Cheiracanthium inclusum*, and *Oxyopes salticus* are just some of the predators that prey on leaf beetle.
- Parasitoids on leaf beetle include the *Crematogaster* sp., *Eupelmus* sp., and *Conura* sp.
- *Nosema* sp. and *Mattesia oryzaephili* are two of the diseases that attack leaf beetles.

DISEASES

Leaf and stem blight, *Macrophomina phaseolina*

Symptoms

- Plants at the five-month mark showed increasing necrosis, followed by leaf dryingand fall.
- Rapid necrosis spread from petioles to stems and ultimately killed off entire branches.

Management

- Treatment with *Pseudomonas fluorescens* decreased the frequency and severity of charcoal rot and enhanced yield.
- *P. fluorescens* production of -1, 3-glucanase significantly inhibited *M. phaseolina.*
- It was also discovered that *Trichoderma harzianum* was useful.

References

Diaz, R., Hibbard, K., Samayoa, A., & Overholt, W. A. (2012). Arthropod community associated with tropical soda apple and natural enemies of Gratiana boliviana (Coleoptera: Chrysomelidae) in Florida. The Florida Entomologist, 95(1), 228-232.

Iriarte, F., Rosskopf, E., Hilf, M., McCollum, G., Albano, J., & Adkins, S. (2007). First report of Macrophomina phaseolina causing leaf and stem blight of tropical soda apple in Florida. Plant Health Progress, 8(1), 40.

25 Stevia

INSECT PESTS

Caterpillar

Damage symptoms

» Caterpillars may feed on tender new growth like leaves and buds.

Management

» Apply Cypermethrin, Deltamethrin, and Quinalphos at a rate of 1.5 to 2.0 ml per litre of water every 15 days.

Whitefly

Damage symptoms

» When plants are brushed or otherwise disturbed, a swarm of these little white flies will rise into the air.

» Whiteflies are tiny flying insects that feed on plant sap by crawling down the undersides of leaves.

» Nymphs and adults alike can cause harm to plants by sucking the juices from tender new growth, which can result in yellowing, spotting, deformity, wilting, and the premature falling of leaves.

» Similarly to honeybees, whiteflies secrete honeydew. This sweet, sticky material coats plants and promotes the development of a black mould. In severe circumstances, this black fungus can prevent photosynthesis from occurring by covering plants.

» In severe cases of whitefly infection, plants may die.

- If plants don't get enough water, they weaken and get sick.

Management

- As a preventative strategy, cover plant beds with reflective paper or plastic mulch.
- Prevent whiteflies from landing on plants by setting out sticky pads in their vicinity.
- To physically remove the insects and their eggs, try spraying the leaves of plants with a high-pressure hose.
- Some people have reported success with using hot pepper wax or insecticidal soap to ward off insects.
- Make an emulsion of mineral oil and soap, or spray Neem oil on the area.
- Take use of yellow sticky traps.
- Use either synthetic pyrethroids like dichlorvos or imidacloprid, or the pesticide dichlorvos.

Aphids

Damage symptoms

- Pestilence-causing sucking insects can be seen feeding on the undersides of leaves and can have a variety of colours, including greenish, red, black, and peach.
- Aphids eat plant juices and sap by boring holes in soft plant tissue like stems and leaves.
- They degrade plant tissue and encourage the growth of sooty mould, which prevents photosynthesis from occurring.

Management

- Use a powerful spray to remove them.
- Set up sticky yellow traps.
- Aphids are repelled by plastic mulch coated with a metalized silver sheen, which reflects sunlight.
- Use a soap that kills insects.
- Insecticides based on Neem oil or mineral oil can be used.
- Use conventional pesticides like Imidacloprid, Malathion, and Acephate, or synthetic pyrethroids like Permethrin.

Cutworms

Damage symptoms

- Cutworms, the larvae of certain moths, spend the daytime underground feeding on decomposing matter before emerging at night to feast on plants.
- The larvae of these insects feed on the plant's stem, killing it in the process.
- Biologically speaking, cutworms are more closely related to caterpillars than worms.

Management

- Carbofuran 3G or Phorate 10G applied to the soil, then irrigated.
- To remove pests, try using a spray made from *Bacillus thuringiensis var. kurstaki* or some baits laced with the bacteria.

Thrips

Damage symptoms

- The wings of thrips, which can be either black or straw in colour, are fringed.
- They feed on the nectar of flowers and eat the leaves, stems, and petals of plants.
- Speckling, discoloration, or silvering of the leaf surface will characterise the plant.
- Many plant diseases are transmitted by thrips.
- Sheets of aluminium foil placed in the spaces between plant rows may deter many thrips.
- Frass, which resembles black varnish, is often seen after a thrips has fed (excrement).

Management

- Pick weeds out of the flowerbed.
- After frost, clean up the bed.
- Insecticides based on Neem oil or mineral oil can be used.
- Set up sticky yellow traps.
- Whitefly activity can be greatly reduced by using plastic mulch with metalized silver sunlight reflecting upper surfaces.

- Use either synthetic pyrethroids like dichlorvos or imidacloprid, or the pesticide dichlorvos.

Red spider mites

Damage symptoms

- The red spider mite consumes the contents of plant leaf cells by puncturing them with its mouthparts.
- Where the cells have perished, the leaves will seem to be finely speckled with brown.
- In addition, adult mites create delicate silk webbing that covers leaf surfaces.
- Depending on how bad the infestation is, this could end up covering the entire plant.

Management

- Insecticides based on Neem oil or mineral oil can be used.
- Eliminate the mites by washing the plants thoroughly.
- Use a spray insecticide containing Dicofol, Dinocarp, Abamectin, Bifenazate, Hexythiazox, or Spiromesifen.

DISEASES

Leaf blight, *Alternaria alternata*

Symptoms

- On the upper surface of the leaves and along the midrib, little circular dots of light brown colour with white to grey centres first developed.
- Some stayed spherical with concentric rings or zones, while others grew irregular and dark brown to grey with time.
- Several individual necrotic patches on severely diseased leaves merged to form larger ones.
- Spots with a circular or radial pattern were more prevalent near the ends of older leaves.
- Leaf spots ranged in size from 2mm to 18mm.
- Stems may wilt if the lesions spread and envelop them.

Favorable conditions

- During periods of high heat, moisture, or humidity, this disease tends to flourish.

Management

- Don't soak the grass or plants.
- Do not touch damp plants, and cut off diseased areas.
- Make sure there's plenty of ventilation.
- Use a 2.0 g/L water spray concentration of Bavistin and Dithane M-45 in rotation.
- In the field, a talc-based formulation of the luminous *Pseudomonas* strain BRL-1 reduced disease severity by 86-87% after being applied to the leaves.
- In addition to lowering the occurrence of leaf spots in greenhouse-grown stevia, *Pseudomonas fluorescens* isolates AUPF6 and AUPF5 promoted greater plant development.

Stem rot, *Sclerotinia sclerotiorum*

Symptoms

- When the plants were about 30 cm tall, dark brown sores appeared on the diseased stems, both above and below ground.
- Plants stunted under dry conditions due to slight lesions on their stems, with the bottom leaves turning black and curling inward.
- When a plant's leaves start to wilt, the damage travels up the plant.
- When the girdling only affected the branches, just some of the leaves began to wilt.
- When the plant's crown and roots become girdled, the entire plant withered and died.
- Under wet conditions, especially after rain, a thin white mycelium mat formed across the bottom of the damaged stems.
- After the plant dies, black sclerotia, measuring between 3.5 and 10.1 mm in length, grow on the plant's crown.

Management

- Raised beds or hills are a prophylactic strategy against fungal diseases, which thrive in humid, damp conditions and standing water. It's also important to water the soil rather than the foliage.

- Stevia should not be planted on sclerotia-infested soil.
- The most effective strategy for preventing sclerotia buildup and lowering sclerotia levels is crop rotation to non-hosts (such as small grains, corn, or sorghum).

Root rot, *Athelia* (*Sclerotium*) *rolfsii*

Symptoms

- First, leaves would turn yellow and droop, followed by plant wilting and the appearance of white cottony mycelial growth around the plant's collar.
- The appearance of white mycelial growth at the base of stem tissue, which then travelled up the stem and into the roots, was a common occurrence connected with decaying tissue, quick stem death, and rotting root decay.
- Brown sclerotia were seen on the affected areas.
- The common sclerotia ranged in shape from round to rectangular and were first white before turning brown.
- Often times, mycelial development, like a white cable, can be seen forming at the base of stem tissue first thing in the morning.

Survival and spread

- Sclerotia, which range in colour from white to brown, are plentiful around the base of stems where they come into contact with the soil.
- Possible field progression via means of transportable machinery.

Management

- Don't grow the same kind of crop in the same place again or even once in a row.
- Uproot diseased plants and throw them away.
- Verify the soil's drainage capabilities.
- Watering the soil with Bavistin at a concentration of 2.0 g/L.
- Topsin M is the best fungicide for eliminating most types of soil-dwelling fungus.
- The effective dose of azoxystrobin was determined to be 380 gm active ingredient per hectare.

Leaf spot, *Septoria steviae*

Symptoms

» It typically occurs as fruit begins to form on the plants.

» Small, round necrotic lesions with grey centres and dark borders first appear on older leaves near the plant's base and then spread upward into the canopy when conditions are right.

» The production of fungal spores causes the darkening of the spots' centres.

» Later in the season, the canopy will be defoliated due to the accumulation of lesions, which will cause necrosis of entire leaves.

» Foliage slowly dies down and fruits are sunscalded.

Favorable conditions

» In densely planted gardens, disease is most prevalent when it rains.

» Lesions trigger the release of conidia from pycnidia, which can be carried by rain splash throughout the season and up the canopy.

» Conditions favourable to disease transmission include heavy rainfall and cold evenings.

Survival and spread

» Inoculum can survive the winter in plant debris and reinfect stevia plants when new stems sprout in the spring.

Management

» At this moment, field cleanliness is the most effective approach to minimise the spread of Septoria leaf spot.

» Picking in September, before the weather and other factors make it ideal for the spread of disease.

» All plant debris must be removed and disposed of.

» When plants are moist, avoid touching them or brushing up against them.

References

Singh, S. D., & Rao, G. P. (2005). Stevia: The herbal sugar of 21 st century. Sugar tech, 7, 17-24.

Bazazo, K. G. I., Ibrahim, A. S. M., & Salem, F. A. (2012). Insect pests of sugar honey leaf, Stevia rebaudiana Bertoni and associated natural enemies in Egypt. Journal of Plant Protection and Pathology, 3(10), 1059-1066.

Martínez, C. (2015). Stevia rebaudiana (Bert.) Bertoni. A review. Cultivos tropicales, 36(Suppl. 1), 5-15.

Fuente, A. L. O. (2001). Survey of pests and diseases of Stevia rebaudiana Bertoni. *Rev. Ciênc. Tecnol. Dir. Invest. UNA*, *1*, 29-33.

26 Basil

INSECT PESTS

Aphid, *Aphis gossypii*

Damage symptoms

- » Adults are dark brown, and their offspring, known as nymphs, are green.
- » They feed on the sap of plants, causing damage.
- » Pest-infested flower buds and blooms will wither and die.
- » Yellowing of leaves, distortion, necrotic patches on leaves, and reduced growth of shoots have all been linked to aphid infestations.
- » Honeydew, a sticky, sugary fluid secreted by aphids, promotes the development of sooty mould on plants.

Management

- » Before planting, make sure the transplants are free of aphids.
- » Aphids can be deterred from plants with the help of reflective mulches, such as silver coloured plastic.
- » The most effective technique of prevention is the use of insecticidal soaps or oils like Neem or Canola oil.
- » Aphid mortality is dramatically increased after being sprayed with 2% Neem/ Pongamia oil.
- » Additionally, 0.05% Dimethoate or 0.05% Oxydemeton Methyl both serve as effective pesticides.

Chinese rose beetle, *Adoretus sinicus*

Damage symptoms

- If the population is high, they will do extensive harm to the basil.
- Holes in the leaves or a lacy pattern where the veins used to be can be evidence of chewing damage.
- Excrete tiny, dark-colored frass particles onto the plant while they feed.

Pest identification

- At oviposition, the tiny elliptical eggs are a dazzling white, but they fade to a drab creamy white over the next few days.
- The larvae look like white, fat, C-shaped grubs with an obvious head and short legs.
- Pupae start out looking yellowish white, but they darken to a brown colour with time. It's covered from head to toe in tiny hairs. The pupa's size ranges from a half an inch to a quarter of an inch.
- The adult beetles are about half an inch in length and have a robust, pale reddish brown coloration. The beetle's body is coated with tiny white hairs that can make it look almost grey.

Management

- Light traps attract the adults, but the females are typically not caught until they have laid their eggs.
- They are also drawn to plants that are under stress or have been injured by beetles, likely due to the presence of gaseous ethylene.
- Use kaolin clay spray or floating row coverings to shield plants from the elements.
- Take out the adults by hand and drown them in soapy water to remove them.
- Beetle populations can be reduced with the use of insecticidal soaps or Neem oil.
- Adult insects can be effectively controlled by using a foliar insecticide.
- During the rainy season, the "green muscardine" fungus, *Metarrhizium anisopliae*, kills a lot of grubs and adults.

Whitefly, *Dialeurodes* sp.

Damage symptoms

» A single, white, elongated egg is placed on the underside of a leaf.

» The nymphs and adults feed on the sap of plants, causing the leaves to curl, wither, and dry out.

» Heavily diseased leaves will develop a black mould and eventually fall off.

Management

» Pests with a soft body, like the whitefly, can be killed by spraying them with insecticidal soap.

» Adults can be managed by spraying them with 0.05% Dichlorvos or 0.0015% Deltamethrin.

» Sprays of 0.05% Triazophos or Methomyl are lethal to nymphs and eggs.

» The use of Endosulfan or Malathion sprays at 0.05% and Carbofuran 3G granules at 2-3 g/plant is suggested for extremely heavy infestations.

Beet armyworm, *Spodoptera exigua*

Damage symptoms

» Feeding on the terminal clusters of host crops, young larvae can do significant harm before they are spotted.

» It's possible for leaves to be skeletonized and devoured almost entirely.

» Very few people really eat plant stems.

» Seedlings are easily damaged, while older plants can often bounce back.

Management

» Kill the larvae by picking them out one by one.

» Remove the weeds and burn the crop debris.

» The parasitic wasp *Frontina archippivora* has been documented feeding on the pest.

» *Bacillus thuringiensis* bacterial pesticides are more effective.

Tinged bug, *Monanthia globulifera*

Damage symptoms

- Damaged areas of leaves turn yellow, curl, and dry as the sap is sucked by prickly black nymphs and lace-winged adults.
- They cause plant death by sucking the juices from young, vulnerable leaves and stems.

Management

- Sprays with either 1% Neem oil or 1% Pongamia oil kill a large number of insects.
- Neem products containing 1000 ppm of Azadirachtin (AZ) at 5 ml/L, or 0.2% Neem kernel oil (NO), sprayed on the leaves every two weeks until the pests are gone.
- Methyl parathion, fenthion, and phenthoate, sprayed at concentrations of 0.05% each, are all effective for pest management.

DISEASES

Damping-off, *Rhizoctonia solani*, *Pythium* sp.

Symptoms

- Both the seeds and their sprouts perish or fail to survive.
- The infection usually enters near the seedling's base, where it causes the stem to shrivel and dry up.
- The plant wilts and dies, either gradually or suddenly.
- A lack of oxygen in the soil might cause the roots to look dark and soggy rather than white.
- The stem's base may develop a water-logged lesion.

Wilt, *Fusarium oxysporum* f. sp. *basilicum*

Symptoms

- When infected plants are between 15 and 30 centimetres tall, they grow normally at first but then become stunted and die unexpectedly.
- Slow development and leaf yellowing in immature shoots are early symptoms of this fungal disease.

- A wilting plant with dieback and browned stems is showing advanced symptoms.
- Roots and stems of infected plants may become discoloured below the bark.
- This symptom can be seen in cross-sections taken from the stem at various lengths.
- The affected parts of the stem will develop a cottony, white to pink growth when exposed to moisture.
- By promptly removing all infected plants and leaves, *Fusarium* can be contained.
- The "Nufar" variety of basil has proven to be resistant. Specialty basils like lemon basil and purple basil have been found to be resistant to the blight.
- Soil solarization combined with dazomet (98G) at 40 g a.i./ m^2 near plants.

Favorable conditions

- Heat and humidity are ideal conditions for the growth of most pathogens.

Survival and spread

- The pathogen is spread through seed and can live for a long time in detritus on the soil's surface.
- The Fusarium wilt pathogen can live in the soil for 8-12 years after it has invaded a field.
- Rapidly dispersed by contact with water, air, or soil.

Management

- Clean seed and new potting soil are the keys to success in maintaining order.
- Seeds can be sterilised by soaking them in a solution of 10% household bleach (1 part bleach to 9 parts water) for 1 minute, then washing them completely with running water before planting.
- The seed needs to be steeped in cold water for four hours, then heated for 20 minutes between 56 to 58 °C.
- The spread of disease can be stopped by removing any infected plants or sections of plants from the garden.
- Don't water basil plants from above.
- Basil should be replaced when planted outside. Sweet basil and other mint

relatives should be avoided in the garden. Although mints won't show any signs of the disease, they may still be infected with it and pass it on from year to year.

Downy mildew, *Peronospora belbahrii*

Symptoms

- Leaves turning yellow with a fuzzy, grey growth on the undersides of the leaves are a telltale sign of downy mildew.
- Leaves are turning yellow.
- The upper part of the leaf may be discoloured.
- Chlorotic spots on leaves rapidly darken to brown in high humidity.

Favorable conditions

- Wet circumstances make downy mildew worse.

Survival and spread

- Seeds from sick plants, infected plant components, and windborne spores from infected basil plants are the most common vectors of spread.

Management

- For healthy basil growth, a well-drained, well-aerated soil is essential.
- Leaves to stay dry for longer, stop watering them from above. Apply drip irrigation.
- All except the least affected plants should be thrown away.
- Actinovate AG foliar spray application following the onset of symptoms.
- To prevent the spread of inoculum from older plants, young ones should be kept separate.

Gray mold, *Botrytis cinerea*

Symptoms

- Colonies of brownish-gray fluff on the stems, leaves, and dead branches of plants.
- Withering and falling leaves on the plant.
- Death of the plant may result from severe lesions on the stem.

Favorable conditions

» Caused by a lack of ventilation and a lot of moisture in the air.

Management

» Take away any plants or leaves that look infected.

» Don't water the lawn from above.

Bacterial leaf spot, *Pseudomonas cichorii*

Symptoms

» Water-soaked brown or black patches, often with jagged edges, on leaves.

» There are streaks on the stems.

Favorable conditions

» The transmission of disease is facilitated by both high humidity and overhead watering.

Management

» The use of inoculated seed and/or disease-free transplants is strongly recommended.

» Field spacing should be kept relatively large to encourage air flow around the plants.

» Eliminate all unhealthy leaves from the plant and the top layer of soil immediately.

References

Garibaldi, A., Gullino, M. L., & Minuto, G. (1997). Diseases of basil and their management. Plant Disease, 81(2), 124-132.

Minuto, A., Bogliolo, A., Minuto, G., Vovlas, N., Troccoli, A., & Scortichini, M. (2009, September). Pests and diseases of sweet basil after methyl bromide phase out: the northern Italian experience. In VII International Symposium on Chemical and Non-Chemical Soil and Substrate Disinfestation 883 (pp. 107-110).

Bernal Areces, B., DeronceléCaigñet, R., & Díaz Pérez, T. (2012). Pests and diseases that affect white basil (Ocimum basilicum) growing under protected conditions. Fitosanidad, 16(2), 87-89.

27 Eucalyptus

INSECT PESTS

Gall wasp, *Leptocybe invasa*

Damage symptoms

- Seedlings, coppiced shoots, and young trees (less than 3 years old) are the primary targets of this wasp's assaults.
- Galls appear as round, shiny growths on the plant, and they are easily distinguished from one another.
- Galls grow to their full size and take on their characteristic hump appearance.
- While the characteristic shine remains, the once-green gall surface gradually turns pink.
- Galls develop on the leaves or stems, and their previously glossy surfaces become dull and a shade of red, light, or neither.
- Galls are easily identified by their emerging holes and their subsequent colour shift from green to brown on leaves and reddish brown on stems.
- Affected seedlings become unusable because their development is stunted.

Management

- Neglected coppice branches provide a breeding ground for gall insects, thus they should be trimmed and burned together with severely damaged plants in the nursery.
- Limiting or stopping the production, distribution, and planting of vulnerable clones is necessary until the issues are contained.

- Parasitoids such as *Aprostocetus* sp. and *Megastigmus* sp. in Australia, and *Ophelimus maskellis* in Israel, are crucial in preventing the spread of *L. invasa* in those countries.
- Insecticides with granules for use in nurseries,
- Set up sticky traps.
- Recognizing clones with resistance.
- Effectiveness was observed following Chlorpyriphos (0.05%) soil treatment, Imidacloprid (0.005%) foliar spray 7 days after soil treatment, Monocrotophos (0.05%) foliar spray 21 days after first foliar spray, and DDVP (0.05%) foliar spray 15 days after second foliar spray.
- Poly-monocultures, as opposed to pure monocultures, of trees help reduce the spread of exotic pests and the depletion of forest resources.

DISEASES

Powdery mildew, *Erysiphe* sp., *Sphaerotheca* sp.

Symptoms

- Powdery mildew, a disease that affects eucalyptus trees, can be identified by the characteristic white powder that appears on the leaves of infected plants.
- It infects eucalyptus, creating spots and deformities on older growth as well as a thick layer of tightly interwoven white mycelium on the surface of leaves and young shoots.
- New limbs may become misshapen, and the tree's growth may be stifled as a result of this disease.
- If the leaves are diseased, they will fall off early.

Survival and spread

- The fungus is able to persist in tough environments like *Erysiphe* and *Sphaerotheca* cleistothecia.
- Conidia are formed in succession on specialised hyphae that emerge from the mycelium's outermost layer, and then are carried by the wind to other hosts, where they might start new infections.

Management

- In order to stop the spread of a disease, it is crucial to spot diseased plants as soon as possible and remove them from the environment.

- Destroying dead leaves can help lessen the risk of an outbreak.
- The use of a fungicide spray, like Benomyl, Chlorothalonil, Triademefon, Maneb, or Zineb.
- Products for commercial biocontrol using *Trichoderma harzianum* T39, *Ampelomyces quisqualis*, *Bacillus*, and *Ulocladium* have been created.

References

Roux, J.*, Coutinho, TA*, Wingfield, MJ* and Bouillet, J. (2000). Diseases of plantation Eucalyptus in the Republic of Congo. South African Journal of Science, 96(8), 454-456.

Wingfield, M. J., Slippers, B., Hurley, B. P., Coutinho, T. A., Wingfield, B. D., & Roux, J. (2008). Eucalypt pests and diseases: growing threats to plantation productivity. Southern Forests: a Journal of Forest Science, 70(2), 139-144.

Old, K. M., Wingfield, M. J., & Yuan, Z. Q. (2003). A manual of diseases of Eucalyptus in South-East Asia. Cifor.

28 Jasmine

INSECT PESTS

Bud worm, *Hendecasis duplifascialis*

Damage symptoms

- The pest is most noticeable once the monsoon season ends.
- Infected *Jasminum sambac* and *J. pubescens* buds are most vulnerable when they are still in the bud stage.
- There are larvae inside the bud or in the flower cluster, feasting on the buds.
- In the early stages, they tunnel into the bud's innermost petals for food and eventually emerge from a circular hole on the tubular region of the corolla to continue feeding and spreading to other buds on the same plant.
- The larvae cause damage to a flower cluster by creating silk and waste tunnels within the cluster, which can prevent flowers from opening and cause flower buds to fall off.
- A rosy hue may develop in the flower buds as well.
- In heavy infestations, the webbing they use for pupation, which occurs either in the soil or on their webs.
- Bushes that have been severely damaged seem charred.

Pest identification

- Round and white when first laid, eggs eventually turn yellow.
- Larvae are a creamy yellow tint with a dark black head and prothoracic region.

- The adult is a tiny, white moth with black spots on the wing margin and wavy lines on the wings themselves; its palpi are black and its proboscis is scaly.

Management

- After the season has over, rake the ground to uncover the pupae.
- If a few infected buds are found during the early stages of an infestation, the contaminated plant components should be collected and destroyed.
- Reducing the insect population by hand by picking larvae and pupae.
- Moth adults can be attracted and killed with a light trap.
- When first instar larvae are spotted, spray the area with Neem oil (1%), Neem cake extract (5%), or Neem seed kernel extract (5%). To be repeated if necessary, wait 10 to 12 days between applications (Nelson et al., 1993).
- The following should be sprayed: Acephate 75 SP at 1.5 g/L, Cypermethrin 25 EC at 1 ml/L, Quinalphos 25 EC at 2 ml/L, Profenphos 50 EC at 1 ml/L, Endosulfan 35 EC at 2 ml/L, Malathion 50 EC at 2 ml/L.
- Protect species of *Perilampus* and *Phanerotoma* and *Mesochrous* that prey on larvae.
- At a concentration of 0.1%, *Bacillus thuringiensis* proved to be an effective pesticide (Sundarababu, 1969). In addition, *B. bassiana* at a spore concentration of 1 x 107 ml1 proved effective.

Web worm, *Nausinoe geometralis*

Damage symptoms

- For *J. sambac*, this is a major problem.
- It is possible to spot eggs on fully developed leaves.
- In the early stages of their lives, newly hatched larvae scrape leaf tissue, while later on, fully grown larvae web leaves and branches together with silken threads.
- Caterpillars skeletonize and dry up leaves because they feed en masse, scraping away the parenchymatous tissues.
- Pupation occurs within the web.
- Extremely damaged shrubs have a scorched appearance.
- The months after the monsoon season are the most destructive. It is peak season for this bug in southern India between the months of July and December.

Pest identification

- The caterpillar is green with brown warts that sprout into fine hairs.
- The adult moth is about the size of a thumb and has wings that are a light brown with white markings.

Management

- The population of insects can be controlled by collecting and destroying webs containing larvae and pupae.
- Cut back the ravaged shrubs.
- Sticky yellow-orange traps at a density of 5 traps per acre can be used to track whitefly populations.
- Moth adults can be attracted and killed with a light trap.
- In the beginning stages of an infestation, spray Neem seed extract (5%) twice every two weeks.
- The use of insecticides like Neem oil or soaps could help control numbers.
- Long-term systemic control can be achieved with 2 ml/L of imidacloprid or 2 ml/L of dimethoate.
- Apply Acetamiprid 20P at 80 gm active ingredient per hectare or Neem oil diluted to 3 ml per litre of water.
- Some research has shown that spraying the web worm with either 0.1% Carbaryl, 0.05% Quinalphos, or 0.01% Fenvalerate is successful.
- Both methyl parathion and phenthoate, at concentrations of 0.05%, are effective in preventing insect infestations.
- The green lacewing, *scymnus*, and *chrysoperla oculata* are only few of the common predators.

Blossom midge, *Cantarinia maculipennis*

Damage symptoms

- Four to five days after oviposition, the maggot causes a purple discoloration of the flower by attacking the petals of *Jasminum sambac* and *J. auriculatum*.
- The inner petals, anthers, and stigma are a favourite meal for these drab white maggots.
- Swelling at the base of the buds is caused by blossom midge maggots, which enter the buds through the corollas.

- The plants get stunted and shrivelled as a result of the infestation.

Pest identification

- Maggots in their mature state are a yellowish orange tint.

Management

- To eliminate pupae, rake the dirt surrounding plants.
- Collect the discoloured flower buds and throw them away.
- It might also be beneficial to set up light traps or sticky traps for this bug.
- Carbofuran 3G applied at 40 g per bush to the soil or Cypermethrin 25 EC applied to the leaves at a rate of 1 ml/L (David and Natarajan, 1998), Triazophos 40 EC applied at a rate of 1.5 ml/L, or Profenphos 50 EC applied at a rate of 1 ml/L every two weeks.
- As a soil treatment, use 250 kg per hectare of Neem cake.
- Neem oil diluted to 2% and sprayed along with 0.05% Teepol during the oviposition phase significantly lessens the harm (David *et al.*, 1992).
- Use either 1% azadirachtin or 5% NSKE as a spray.

Gallery worm, *Elasmopalus jasminophagus*

Damage symptoms

- Every species of *Jasminum* is vulnerable to this pest.
- It's a major problem for *J. auriculatum* as it causes the plant to web.
- The caterpillar builds silken and excremental matter tunnels on the tips of leaves, shoots, and flower heads, where it then feeds.
- The silky web has faeces stuck to it.
- The galleries are where the pupation process takes place.

Pest identification

- The caterpillar is green overall, but has a reddish brown head and brown stripes along either side.
- The moth is long and skinny, with dark grey forewings and lighter grey rear wings.

Management

- Gather the broken blossoms and silken tubes and destroy them.
- Moth adults can be attracted and killed with a light trap.
- Application rates of 2 ml/L of NSKE 5%, Endosulfan 35 EC, and Malathion 50 EC are recommended.
- Bushes need regular, meticulous pruning to keep them healthy and looking their best.

Shoot borer, *Sycophila* sp.

Damage symptoms

- Pest damage to *J. grandiflorum* has been documented.
- The larvae burrow into the shoots, eat and pupate within them, and then emerge as adults.
- Damaged shoots have a pattern of holes in them.

Management

- Destroy and prune infected shoots.
- Shoot borer infestations can be stopped by applying Carbofuran at a rate of 1.0 kg a.i. /ha or spraying with 0.06% Triazophos or 0.05% Chlorpyriphos every two weeks.

Leaf roller, *Glyphodes unionalis*

Damage symptoms

- The caterpillar feeds by rolling the leaves around on its body.

Pest identification

- The caterpillar's colour is green.
- The costal border of the white moth's adult wings has brown lines.

Management

- Collect the ruined leaves and dispose of them along with the larvae.
- Bushes need regular, meticulous pruning to keep them healthy and looking their best.
- Moth adults can be attracted and killed with a light trap.

- Use a spray containing either 5% NSKE, 2% Endosulfan 35 EC, or 2% malathion 50 EC.

Tingid bug, *Corythauma ayyari*

Damage symptoms

- Among the many things that bother *J. sambac*, this is one of the worst.
- Adults and juveniles alike might be found sucking the juices off the leaves.
- Infected leaves dry up and fall off after turning yellow.

Pest identification

- The adults' lacewings are white in colour.

Management

- Infested leaves and all stages of the insects should be collected and disposed of.
- Methyl Demeton 25 EC 2 ml/L, Dimethoate 30 EC 2 ml/L, or NSKE 5% can be sprayed.

Jasmine bug, *Antestia cruciata*

Damage symptoms

- Both nymphs and adults feed on the sap of young plants, stunting their growth and thwarting bloom development.

Pest identification

- Nymph has a spherical body and is a dark brownish black colour.
- The wings of an adult are marked with orange and white, and the body is a dark brown shape resembling a shield.

Management

- Apply a 2 ml/L monocrotophos spray.

Flower thrips, *Thrips orientalis*

Damage symptoms

- Adults and juveniles alike are attacking the blossoms.
- Petals on flowers often have brown spots.

Pest identification

- Nymphs are often golden in colour.
- An adult has a dark skin tone.

Management

- Acephate (0.1%), Imidacloprid (0.01%), and Azadirachtin (1%) in rotation was proven to be effective.
- Methyl Demeton 25 EC 2 ml/L, Dimethoate 30 EC 2 ml/L, or NSKE 5% can be sprayed.

Gall mite, *Aceria jasmini*

Damage symptoms

- Erenium, felty velvet-like hairy outgrowths caused by the pest's colonisation and feeding, cause deformation of vegetative and floral sections, which in turn affects crop growth and output.
- Both *J. auriculatum* and *J. pubescens* are severely plagued by gall mite.
- It feeds on the plant's epidermis, young stems, and budding organs.
- As a result, the plant's growth slows down and its flowering capacity diminishes.
- This insect has a significant impact on crop yield when grown for commercial purposes.

Pest identification

- Females have two sets of legs and are cylinder-shaped. approximately 150-160 in length and 44 in thickness.

Management

- During the dormant season, prune the affected areas and then spray them with 2% Neem oil or 2% Pongamia oil. After three weeks, try it again.
- To effectively control gall mites, spray them with 0.3% Wettable Sulfur 80 WP at 3 g/L or 0.06% Triazophos 40 EC at 1.5 ml/L at the start of the monsoon and again 14 days later. Repeat this process two or three times.
- Galls caused by the mite can be reduced with the help of sprays containing either 0.05% Ethion or 0.01% Fenproparthin and 0.02% Dicofol.
- Applying a spray of 0.05% Dicofol or 0.3% Wettable Sulfur is an effective method of preventing spider mite infestations and preventing the development of spider mite eggs.

Red spider mite, *Tetranychus cinnabarinus*

Damage symptoms

- The silken webs conceal feeding mouths as they feast on the undersides of leaves.
- Leaves' upper surfaces develop yellow spots, which eventually splotch, brown, and fall off.
- Last of the affected leaves to fall off.
- The effects on development and blooming are negative.

Pest identification

- Adults and juveniles both have red colouring.
- White, spherical eggs are placed on the underside of the leaves.

Management

- Take out and burn the offending pieces.
- Apply a solution of Wettable Sulfur 50 WP 2 g/L or Dicofol 18.5 EC 1.5 ml/L.
- Two ml per litre of Endosulfan spray.

DISEASES

Collar rot and root rot, *Sclerotium rolfsii*

Symptoms

- Disease affects plants at all developmental stages. The older leaves turn yellow first, followed by the younger ones, and then the whole plant dies.
- The root has a dark stain on it.
- Mycelia, which look like white strands, and sclerotia, which like mustard seeds, can be visible on the surface of the infected tissues and the stem itself.
- New leaves become wilted and twigs dry out from the top down.
- There are dry spots among the plants.

Management

- Watering the soil with either a 1% Bordeaux mixture or 0.3% Copper oxychloride solution helped lessen the impact of the wilt disease.

- Trifloxystrobin + Tebuconazole 0.75 g/L or Difenoconazole 0.5 ml/L applied to the soil.
- Plants should be treated with *Pseudomonas fluorescens* at 25 g/m2 in the soil and 5 g/L in the foliage once a month after planting.
- A Bordeaux combination of 1% copper oxychloride or 0.25 percent copper oxychloride can be sprayed on the soil.
- The addition of *Trichoderma viride* to a high-volume treatment of FYM (10 g + 100 g FYM/plant) is also beneficial.

Leaf spot, *Cercospora jasminicola*

Symptoms

- Symptoms manifest mostly on leaves as necrotic patches that range in shape from circular to oval and have dark brown to reddish edges.
- Later, the dots spread across wider sections of the leaves and take on an uneven shape.
- Dark grey development of the infection is visible beneath the dots.
- Heavily defoliated plants are a common result of a severe disease.
- It's not uncommon to spot symptoms on the stems.

Management

- Spraying with 1% Bordeaux mixture or 0.1% Benomyl or Carbendazim or 0.25% Mancozeb at 15-20 day intervals will help keep the disease at bay.

Rust, *Uromyces hobsoni*

Symptoms

- Certain species of Jasminum are particularly susceptible to this disease, and it is most widespread in South India.
- All of the plant's blooms and other aerial sections show symptoms.
- Spots of a yellowish-orange hue first appear on leaves, young twigs, and flower buds.
- All of a sudden, these areas are really big and puffy.
- Multiple cup-like projections, both orange and yellow in colour, hold spores of the fungus, which may be seen on the underside of the enlarged area.

- Young diseased stems and leaves grow grossly enlarged and misshapen.
- Areas that previously displayed orange, yellow, cup-like projections at the start of the season eventually acquire dark brown, often black, bloated formations (called telial galls) towards the end of the season.

Management

- To lessen the inoculum for the following season, the infected leaves and other plant parts should be gathered and destroyed.
- Spraying 0.2% Mancozeb or Zineb can help slow the spread of the disease.

Leaf blight, *Alternaria jasmini, A. alternata*

Symptoms

- Identified by the rapid proliferation of reddish-brown spots on the leaf's upper surface during the wet season.
- In wet weather, the spots spread over a wider area, eventually producing blighting of the leaves.
- Curling and drying out occurs around the leaf margins of diseased plants.
- The lesions display a pattern of concentric circles.
- Flowers, petioles, and stems are all susceptible to the disease.
- Shoots, vegetative buds, and young branches wither and die when an infection is severe.
- Infected plants have much less bloom yield.

Favorable conditions

- Winter is when the sickness is at its worst (Oct - Dec). It's not uncommon for the disease to be detected in February in some regions.
- Conditions favourable to the spread of disease include persistent drizzle, a relatively cool temperature, and high humidity.

Management

- Raking and dumping of leaves.
- Copper oxychloride 0.25 percent, Mancozeb 0.25 percent, or azoxystrobin 0.1% can be sprayed to kill the bugs.

References

Pirithiraj, U., Soundararajan, R. P., Gailce Leo Justin, C., & Lakshmanan, V. (2020). Influence of abiotic factors on major insect and mite pests of jasmine, Jasminum sambac L. Journal of Entomology and Zoology studies, 8(3), 680-684.

Ashrith, K. N., & Hegde, J. N. (2020). Insect Pests of Jasmine and Their Management. In Advances in Pest Management in Commercial Flowers (pp. 103-118). Apple Academic Press.

Khan, M., Lanjar, A. G., Chang, B. H., Bukero, A., Rajput, A., Magsi, F. H., ... & Chang, A. H. (2019). Insect Pests Associated With Ornamental Plants: Insect Pests of Ornamental Plants. Biological Sciences-PJSIR, 62(3), 188-194.

Kamala, I. M., & Kennedy, J. (2017). Survey on the prevalence of jasmine leaf web worm, Nausinoe geometralis and its natural enemies in Tamil Nadu. Journal of Entomology and Zoology Studies, 5(6), 409-414.

Sudheera, Y., Vishnu Vardhan, G. P., Hema, M., Krishna Reddy, M., & Sreenivasulu, P. (2014). Characterization of a potyvirus associated with yellow mosaic disease of jasmine (Jasminum sambac L.) in Andhra Pradesh, India. Virus disease, 25, 394-397.

Wikee, S., Cai, L., Pairin, N., McKenzie, E. H., Su, Y. Y., Chukeatirote, E., ... & Hyde, K. D. (2011). Colletotrichum species from Jasmine (Jasminum sambac). Fungal Diversity, 46, 171-182.

Al-Zadjali, A. D., Natsuaki, T., & Okuda, S. (2007). Detection, identification and molecular characterization of a phytoplasma associated with Arabian jasmine (Jasminum sambac L.) witches' broom in Oman. Journal of Phytopathology, 155(4), 211-219.

29 Lemon Grass

INSECT PESTS

Aphid, *Macrosiphum muscanthii*

Damage symptoms

- Aphid infestations typically occur on young, fragile branches.
- The nymphs cause the leaves to yellow and the vulnerable shoots to shrink becausethey feed on the plant's sap.
- They wreak havoc on the infected plant, causing deformities and slow growth.

Management

- Aphids can be effectively killed with two sprays of 1% Neem/Pongamia oil separated by 10 days. Aphid damage can be reduced by 0.05% using either dimethoate or oxademeton methyl or profenphos.
- If necessary, spray some Neemark (2%).

DISEASES

Rust, *Puccinia nakanshikii*

Symptoms

- Mild yellow patches on the plant life.
- Leaf veins are discoloured a dull brown, and there are brown elongated streaks onthe leaves.
- Underside of leaves covered in dark brown pustules.
- The plant's leaves and life are over.

- Distinct, linear, cinnamon brown uredinia on the abaxial surfaces of mildly andhighly infected leaves are accompanied by chlorotic streaks.
- Extensive necrosis coincides with consolidated uredinia, indicating a severe infection.
- On badly infected leaves, telia can also be found.
- There is no known host for the aecial stage.

Survival and spread

- In the absence of a host plant, spores can persist on decomposing lemon grass.
- Wind, rain, and splashing water all play a role in the spread of a disease.

Management

- Make sure the plants are disease-free.
- Use mulches and fertiliser to encourage rapid plant development.
- To recover and thrive prune off any infected areas.
- Not water from above.
- Apply a fungicide spray, such as Mancozeb 0.3% or Zineb 0.3%, that works oncontact.
- The hyperparasite *Darluca* sp. feeds on rust.

References

Skaria, B. P., Joy, P. P., Mathew, S., & Mathew, G. (2006). Lemongrass. In Handbook of herbs and spices (pp. 400-419). Woodhead Publishing.

Joy, P. P., Skaria, B. P., Mathew, S., Mathew, G., Joseph, A., & Sreevidya, P. P. (2006). Lemongrass. Ind. J. Arecanut Spices Medicin. Plants, 2, 55-64.

30 Mint

INSECT PESTS

Semilooper, *Thysanoplusia orichalcea*

Damage symptoms

- Injury is caused by the first and second instar larvae scraping the underside of the leaves.
- The damage caused by the third and fourth instar larvae, which feed voraciously down the sides of leaves, can reach over 60% in extreme infestations.
- Based on the maximal survival and growth index, Japanese mint emerges as the superior host.
- The most productive time period was the second week of April to the end of May.
- Mentha arvensis attracted the most moths, followed by *M. spicata*, *M. citrata*, *M. piperita* var. citrata, and *M. piperata*.
- Honeydew, a sticky, sugary fluid secreted by aphids, promotes the development of sooty mould on plants.

Management

- Before planting, make sure the transplants are free of aphids.
- Pruning out the infestation is one way to maintain order.
- Aphids can be deterred from plants with the help of reflective mulches, such as silver coloured plastic.

- Foliage damage caused by the semilooper can be reduced by using NSKE (4% or 1% Neem oil) or Neemark (0.25%).
- The most effective technique of prevention is the use of insecticidal soaps or oils like Neem or Canola oil.
- Methyl parathion, quinalphos, and fenvalerate sprayed at 0.05%, 0.01%, and 0.01%, respectively, achieve good control.
- Parasitoids such as *Cotesia ruficus* and *C. glomerata* have been spotted in Punjab feeding on *T. orichalcea* larvae found in mint.
- Heavy mortality of larvae and pupae was also caused by *Bacillus cereus* var. *mycoides*.
- Also effective are treatments with the *T. orichalcea* nuclear polyhedrosis virus (300 LE/ha) or *Bacillus thuringiensis* (500 ml/ha).

Flea beetle, *Longitarsus waterhousei*

Damage symptoms

- Mint oil production is hindered by flea beetle larvae eating the roots of the plants.
- Mint plants that have been attacked by this pest have a poor regeneration rate in the spring, and their new growth is patchy and has a rusty colour.
- In order to survive, mint larvae tunnel into underground storage organs called rhizomes, which they feed on as they mature.
- When adults eat mint leaves, such holes become characteristic of the species.

Survival

- Flea beetle eggs overwinter in the soil around the mint plant's crown.

Management

- It's important to use rhizomes that have been verified to be flea beetle-free.
- Use entomopathogenic nematodes (BioVector) and keep the soil moist.
- Control adult beetles by using a foliar insecticide (such as Malathion, Lannate, or Actara) about the middle of July, before they have a chance to lay further eggs.

Mint beetle, *Chrysolina menthastri*

Damage symptoms

» The mint leaf beetle, with its stunning metallic green colouring, is found primarily in wet meadows and at water's edge.

» Its summertime diet consists primarily of the leaves of several mint plants.

» Later in the summer, a blackish, squishy-looking grub may be spotted consuming mint leaves.

Management

» This pest can be managed with timely spraying.

» Grab the old and the young and put an end to them.

Bud mite, *Floridotarsonemus* spp.

Damage symptoms

» Internodes at the tips are cut shorter.

» New leaf curling up.

» the apical bud twists or puckers.

» Damage caused by spider mites looks as tiny silver or dry spots on the upper surface of the leaf.

» Leaves that have been severely injured become brown and fall off.

» Essential oil production may drop drastically (by 60-80 percent) as a result.

Pest identification

» Eggs range in hue from transparent to a milky white, have an oblong form, and are roughly 75% the size of an adult female.

» The mite's immobile larvae have six legs and are capable of movement.

» The mature stages of these mites, which have eight legs, range in colour from transparent to pale yellow and eventually darken to a golden amber. Women are typically bigger and rounder than their male counterparts.

Favorable conditions

» Temperature increases in the spring could hasten the mite population explosion.

- A generation of mites can be finished in as little as a week when temperatures range from 21 to 30 °C.

Survival

- The adult female mint bud mite hibernates beneath the earth on mint stolons or plowed-down mint debris.

Management

- To apply Kelthane, Comite/Omite, and Fujimite, spray the appropriate chemicals.
- Rotate your crops every two or three years.

Red spider mite, *Tetranychus urticae*

Damage symptoms

- Spider mites cause harm by puncturing microscopic holes in leaf cells, and they flourish in hot, dry environments.
- Spotted darkening on leaves and a fine webbing reminiscent of a spider's web are telltale signs of their presence.
- Leaves with tiny yellow dots.
- Sometimes the leaves can look almost bronzed.
- As if the leaves were webbed over.
- Spot mites on the underside of leaves or in spider webs.
- The plant's leaves may become yellow and fall off.

Favorable conditions

- In a dusty environment, spider mites can flourish.
- Plants that are water stressed are easier to damage.

Survival

- Mature female mites hibernate in the bottom of mint stems and in the soil.

Management

- Simply soaking the plants in water will do the trick.
- Plants can benefit from using insecticides such as Neem oil or insecticidal soap.

- Abamectin, an acaricide, can be sprayed.
- Try the incredibly effective garlic water or hydrogen peroxide treatment.
- When about 20% of leaves in a sample have fewer than one spider mite, release spider mite predator *Neoseiulus fallacis* at a rate of 5,000 individuals per hectare.

DISEASES

Powdery mildew, *Erysiphe cichoracearum*

Losses

- Defoliation caused by the disease can cut essential oil production by as much as 20%.

Symptoms

- Before the disease enters its powdery phase, little chlorotic spots may emerge on the surface, and the underside of the affected area will become discoloured and brown.
- A few stages later, the upper surfaces of the surviving leaves develop white powdery spots.
- Banding on the stem and branches indicates extensive defoliation.
- These white areas eventually darken to a purple-black and form several perithecia.

Survival and spread

- Primary spread occurs via resting spores, while secondary spread occurs via airborne spores.

Favorable conditions

- When airflow is restricted and temperatures are high, diseases flourish (warm and humid weather).

Management

- It's important to remove diseased plants and avoid overhead watering.
- Wet-Sulfur Dusting.
- The condition can be properly managed with the use of Karathane.

Wilt, *Verticillium* spp.

Symptoms

- Dwarfing, lopsided branch growth, leaf etiolation, wilting, and mortality are all common manifestations of this disease.
- Leaves towards the top of the plant curl and twist first.
- The plant's leaves are clustered together near the crown.
- Stunted and with yellow, crimson, or bronzed leaves are symptoms of plant infection.
- First the lower leaves, then the whole of the plant above ground, die.
- In the case of flowering or other pressures, stem or plant death may occur too quickly for these symptoms to be noticed.

Favorable conditions

- Wet soils and temperatures between 21 and 27 °C are ideal for the growth of Verticillium wilt.
- Host plant root exudates promote the germination of microscopic sclerotia.

Survival and spread

- Soil, roots, and stolons can all be utilised to propagate the infection, as can infected seed.
- Without a plant host, a pathogen can survive in soil for a very long time.
- Infected seedlings and other planting supplies can spread the disease over great distances if they are shipped.
- Pathogens spread predominantly through soil and the movement of soil and water.

Management

- Summer time soil inversion.
- A focus on using only wholesome seeds and soil.
- The highest yield of healthy mint rhizomes is obtained with heat treatment at 47°C for 55 minutes or 48°C for 35 minutes.

Stolon rot, *Macrophomina phaseoli* (*Rhizoctonia bataticola*)

Symptoms

- Lesions on underground stolons start off pinkish brown and progress to dark brown or black as the disease advances.
- Leaves turning yellow and plant development slowing down are the most prominent early indications.
- The plants' above-ground parts wilt and die in the pre-mortem phases.
- These spots grow in size and eventually cause the soft disintegration of the affected area or the entire stolon.

Management

- Plants are used that are in good health.
- Summer time ploughing is done to depth.
- Keeping water usage down.
- Damaged vegetation is to be uprooted and burned.
- Rotation of crops, essentially.
- When the stolons are treated with Zineb, Mancozeb, or Captan prior to planting, the disease can be managed well.
- Disease control can also be aided by adding *Trichoderma* sp. to the soil.

Rust, *Puccinia menthae*

Symptoms

- The undersides of leaves develop brown, round pustules, which are quickly followed by leaf withering and eventual defoliation.
- In the spring, yellow uredospores can be found in reddish blisters on new shoots, while in the fall, black teliospores can be found on stems and regrowth.
- A swollen telium is a hallmark of the telial stage.
- Internode elongation and chlorosis in the context of a swollen shoot are symptoms of a systemic disease. The primary symptom of this disease is leaf loss, which limits mint oil production.

Favorable conditions

» High relative humidity in dry conditions.

Survival and spread

» *Puccinia menthae* survives the winter on mint weeds and mint that has grown wild.

» Exists eternally as uredospores on the host plant's runner and dead leaves and stems.

» Later-stage, thick-walled teliospores may endure harsh environments.

» It can travel through the air and other propagation mediums.

» It is the wind that carries the orange spore pustules to other plants when they have matured and burst open in June or July.

Management

» Most importantly, this sickness can be avoided by the use of materials that have not been exposed to the disease.

» Planting material sterilisation at 44 °C for 10 minutes.

» To stop the spread of the disease, infected plants and rhizomes must be destroyed.

» For effective rust prevention, rabbining (flaming) must take place.

» Crop rotation is an effective method for managing the disease.

» It's possible that rust damage can be mitigated by harvesting early (before the leaves fall off due to rust).

» Protection from rust disease can be achieved through the use of fungicides such as Mancozeb, Propiconazole, Diclobutazol, and Nickel chloride.

» Chlorothalonil, Pycroclostrobin, or Propiconazole sprays should be applied when plants are between 5 and 10 cm tall or when environmental circumstances are conducive for disease development.

» Absolutely pristine c.v. The genotypes MAH-3, S-10-11-45, and S-13-2-125 were only moderately resistant to rust fungus, but the Himalaya, Kalka, and Damroo varieties were very resistant. Kalka was able to resist rust, leaf spot, and leaf blight better than other varieties.

Black stem rot, *Phoma strasseri*

Symptoms

- Cankers and lesions ranging from black to dark brown emerge on the stems and rhizomes.
- The girdling effect of cankers on a plant's stem can result in the wilting and death of the plant tissue above the point of infection.
- Cankers girdling the stem at or near the soil line cause the foliage of infected plants in the field to turn red, stunt, and eventually die.
- Lesions on rhizomes become black and wet.
- Infections frequently spread from rhizomes to emerging new shoots, and lesions on rhizomes can consolidate and progress to cause general necrosis and rot.

Favorable conditions

- The conditions for the spread of disease are ideal when the soil is warm and damp.
- Optimal conditions for the spread of disease include temperatures between 26 and 32 °C, soil wetness between 30 and 60 percent, and a soil pH over 6.6.

Survival and spread

- Black or brownish resting structures (sclerotia) or resting fungal mycelium are how this pathogen thrives on crop debris and in soil.

Management

- Getting hold of robust rhizomes and then planting them.
- Preventative measures include treating the stolons with a 0.25% solution of Captan or a 0.1% solution of Benlate for 2 to 3 minutes before to planting.
- Aiding plant development by dietary supplementation and cultural techniques.
- Avoid over-irrigating young plants, as this increases their susceptibility to disease.
- Before and during planting, take care not to damage the rhizomes, and throughout the course of culture, prevent the stems from being punctured.

» Rice, wheat, and mint should be rotated every three years.

» The mint shoot tips that were treated with a fungicide soak were effective in preventing the spread of the fungus.

Begomo virus

Symptoms

» Foliar rugosity, yellow vein, leaf yellowing, light yellow mosaic, upward curling, crinkling, and slowed growth can all be symptoms, all of which contribute to a significant drop in yield.

» Cause the leaves to curl and cup noticeably, reduce in size, and slow the plant's growth.

» There was a complete loss of biomass that might be used to make herbs when plants were infected at a very early stage or after transplantation.

Spread

» Naturally affected *M. spicata* cv. Viridis plants served as vectors for whiteflies to spread the disease to healthy seedlings (*Bemisia tabaci*).

Management

» Plant or cultivate seeds that are resistant to viruses and whiteflies.

» When whitefly populations are low, removing the early symptomatic plants may be helpful in limiting the spread of viruses.

» Use black (non-metalized) mulch down the middle of UV-reflective (aluminium) mulch.

» When four samples were taken from the field between March and May, the Kushal variety had the lowest level of Begomo virus infection.

» Recent research indicates that pymetrozine (Fulfill) can inhibit viral replication.

References

Kalra, A., Singh, H. B., Pandey, R., Samad, A., Patra, N. K., & Kumar, S. (2005). Diseases in mint: causal organisms, distribution, and control measures. Journal of herbs, spices & medicinal plants, 11(1-2), 71-91.

Rung, A., Halbert, S. E., Ziesk, D. C., & Gill, R. J. (2009). 0088. A Leafhopper Pest of Plants in the Mint Family, Eupteryx decemnotata Rey (Hemiptera:

Auchenorrhyncha: Cicadellidae), Ligurian Leafhopper, New to North America. Insecta Mundi, 1-4.

Verma, J. S. (2006). Insect pest problem in medicinal plants-a review. Agricultural Reviews, 27(2), 130-136.

Singh, A., & Singh, H. B. (2004). Control of collar rot in mint (Mentha spp.) caused by Sclerotium rolfsii using biological means. *Current science*, 362-366.

Suganthy, M., & Mani, M. (2022). Pests and Their Management in Aromatic Plants: (Basil, Davana, Jasmine, Lemongrass, Marjoram, Mint, Patchouli, Rosemary, Sage and Thyme). Trends in Horticultural Entomology, 1305-1329.

31 Patchouli

INSECT PESTS

Leaf caterpillars, *Pronomis profusalis*

Damage symptoms

- Webbed by the larvae of *Pronomis profusalis*, the leaves of young plants are consumed until just the veins remain.
- The caterpillars' feeding habits stunt the development of the plant.

Management

- The amount of pests can be reduced by using clippers and burning rolled up leaves and spider webs.
- Fenvalerate 0.01% followed by Quinalphos or Methyl Parathion 0.05% sprayed twice or thrice is advised for pest management.
- Catch an infestation early on, a spray made of 5% Neem seed kernel extract and 5% Teepol can be very effective.

Pentatomid bug, *Nezara viridula*

Damage symptoms

- Nymphs and adults of the same colour green feed on the sap of young, fragile leaves and shoots.
- The symptoms of attack include deformed leaves and a reduction in the plant's growth rate.

Management

- If egg masses are collected and destroyed, no harm will come of it.
- To prevent damage from insects, spray the plants with a 0.1% solution of Carbaryl or Malathion, or a 0.05% solution of Fenthion.

Leaf roller, *Pachyzancla (Psara) stultalis*

Damage symptoms

- Caterpillars cause leaf damage by rolling the leaves and feeding on the tissue inside.
- The use of leaf rollers reduces oil quality and causes damage.

Management

- Throw away the curled up leaves.
- Teepol (0.1%) can be used as a spray for either Neem oil (3%) or Pongamia oil (3%).
- To eliminate *P. stultalis* larvae, use cinnamon oil at a concentration of 40%.

Mirid bug, *Pachypeltis* sp.

Damage symptoms

- Tender leaves and young shoots are consumed by the nymphs and adults, causing necrotic interveinal patches that eventually form punctures.
- Herbage output was shown to be diminished in cases of severe infestation.

Management

- Research indicates that the Java cultivar is the most vulnerable, while the Singapore variety is the least.

DISEASES

Fusarium wilt, *Fusarium solani*

Symptoms

- This disease is most common during the kharif harvest season.
- The first visible symptom was a yellowing of the elder leaves, which quickly spread to the younger and older leaves at the tips and midsection of the plant.
- Parts of plants that are still green are dried.

- There was plenty of water in the soil, yet the plants look like they're dying of thirst.
- The stem and roots turned a dark brown to black colour, while the vascular tissue turned a dark brown to red tone.
- It was found that the surface of the secondary roots was severely deteriorating.
- Plants that have reached maturity may have a blackening of their roots around the collar.
- Plants that were affected with the disease withered to the point where they could be readily plucked up from the ground.
- The plant as a whole dried out and died early in the end.

Management

- Neem oil was effective in the field against *F. solani.*
- Carbendazim 1%, Carbendazim 1% + Mancozeb 1%, Carboxin 1% + Thiram 1%, Propiconazole 1%, and Mancozeb 0.2% soil soaking.
- Seed treatments with *Bacillus subtilis* and *Trichoderma harzianum* dramatically decreased pathogen and wilt colonisation of seeds.
- There is hope for wilt management by the combination of Carbendazim and *T. harzianum.*
- The lowest disease incidence and highest herbage yield were achieved through a combination of seedling root dipping and soil drenching with 0.1% Carbendazim, 100 g/plant of Biofor-pf-2 (a commercial bioformulation of *Trichoderma harzianum* and *Pseudomonas fluorescens* derived from Vermicompost), and 50 g/plant of mustard oil cake.

Root rot, *Rhizoctonia* spp.

Symptoms

- Plants suddenly wilt as a result of the disease.
- The collar area and the main roots of infected plants turn a dark brown colour.
- When the soil is wet or damp, the disease becomes extremely destructive.

Favorable conditions

- Warm, damp sand is particularly vulnerable to destruction.

Management

- It's crucial to instal proper drainage systems.
- Applying 2 g/L of copper oxychloride to the ground.
- At the roots of affected plants, drenching the soil with 0.2% PCNB (Brassicol) helps to reduce the incidence of disease.
- Disease was halted thanks to a combination of *Gliocladium virens* and *Glomus aggregatum*.

Leaf blight, *Alternaria alternata*

Symptoms

- Symptoms of leaf blight manifested as dark discolouration, starting at the leaf tips of mature plants.
- Marginal necrosis developed as the disease spread to the leaf's outer edges.
- The patches ranged in shape and size from subcircular to irregular, and were a light brown tint.
- The plant's charred appearance, visible even from a distance, made it easy to spot when conditions were optimal.

Management

- More than half of the sickness was stopped by spraying Ferbam, Mancozeb (Dithane M-45), and Captan.

Sclerotium wilt

Symptoms

- Early signs included water-soaked lesions in the neck area, followed by a general darkening of the affected area.
- Eventually, the leaves wilted and fell off.
- Small, light to dark brown sclerotia developed on the infected part of the stem, which was covered in white, fan-shaped mycelial strands.
- Initially, the sclerotial bodies were white, but as they aged, they became brown.
- The plant eventually withered and died.
- The dead leaves also have sclerotial bodies formed on them.

Management

- Plague rates can be lowered with deep summer ploughing and good drainage.
- For this soil-borne pathogen, Topsin M proved to be the most efficient control method.

Yellow mosaic disease, *Yellow Mosaic Virus (Potyvirus)*

Symptoms

- When temperatures rise, typically between April and June, a characteristic golden mosaic forms.
- During the colder months, these symptoms often go undetected.
- Symptoms typically include yellow mosaic disease without leaf distortion, pale green mosaic disease with leaf malformation, severe mosaic, thicker leaves, or stunted and retarded plant growth.
- Sap and aphids were the vectors for the spread of Patchouli Yellow Mosaic Virus (PaYMV).
- The virus caused a characteristic line pattern on the leaves of *Nicotiana benthamiana*, the plant used for virus diagnosis.
- The isolated virus particles had a curved rod form and measured about 805 nm in length after purification.

Management

- At a concentration of 0.05%, Dimethoate stops the aphid vector and, in turn, insect transmission.
- The apical meristem cultivation method may be used to cultivate patchouli plants that are immune to viruses.

References

Gahukar, R. T. (2018). Management of pests and diseases of important tropical/subtropical medicinal and aromatic plants: A review. Journal of Applied Research on Medicinal and Aromatic Plants, 9, 1-18.

Bora, P., & Bora, L. C. (2020). Disease management in horticulture crops through microbial interventions: An overview. Indian Journal Of Agricultural Sciences, 90(8), 1389-1396.

32 Scented Geranium

INSECT PESTS

Whitefly, *Trialeurodes vaporariorum*

Damage symptoms

- These insects are so small that they may fit in the palm of hand, and they get their nutrition from plant sap, which they then excrete as honeydew.
- On these deposits, a black, sooty mould begins to thrive.
- The adults are particularly difficult to kill with insecticides due to their waxy outer covering, which acts as a repellent against water.

Pest identification

- Adults can be identified by their moth-like appearance; they are white in colour and are only 1-2 mm in length, and they reside on the underside of leaves.

Management

- Getting rid of the diseased leaves and burning them.
- A reflective mulch or aluminium foil can keep whiteflies away.
- Using yellow sticky traps can help with whitefly control.
- Detergent in diluted form can be sprayed on the waxy surface to help remove it (a few drops in 1 pint of water). The insect will lose its protective coating and eventually die as a result.
- To combat whitefly, use a pyrethrum spray, which discourages predator insects.

- As a biological control measure, the parasitic wasp *Encarsia* sp., which deposits its eggs inside the eggs of the whitefly, might be introduced to affected areas.
- Whitefly numbers can be reduced through the use of insecticidal soaps or Neem oil sprays.

Aphids, *Myzus persicae, Aphis gossypii*

Damage symptoms

- Aphids are tiny winged insects that resemble whiteflies but are much simpler to remove.
- The first indicator will be creased leaves, followed by the discovery of an aphid colony on the leaf underside.
- Pear-shaped, small insects that feed on plant liquids and can transfer viruses.

Management

- Knocking them off the plants with a forceful stream of water is an effective method of pest management.
- Spraying with horticultural oil, insecticidal soap, or malathion.
- Putting out a call for parasitic wasps, ladybirds, green lacewings, and aphid midges, among other useful insects.
- Spraying with Pyrethrums or another insecticide may be required if first aphid attacks are not successfully repelled (Mavrik or Confidor).

Bud worm, *Heliothis (Helicoverpa) virescens*

Damage symptoms

- Caterpillars of various sizes make holes in flowers by tunnelling through their buds, petals, and leaves.
- Stems become weak and susceptible to breaking as a result of the harm caused by the tiny larvae when they tunnel through them.
- In about one month, a larva can consume enough food to reach sexual maturity.
- In just a single year, it finishes off two generations.

Pest identification

» In part due to the plant they are feeding on, bud worm larvae range in colour from dark brown to red or even green, and are striped.

» The wings of an adult budworm moth are a pale green and are banded with lighter colours over a distance of 2.50-3.75 mm.

Survival and spread

» The budworm pupates in earthen cells a few inches below the soil's surface, where it spends the winter.

» Pupae do not overwinter well when temperatures fall below 20 °F in the winter.

» In most cases, adulthood begins in late spring.

Management

» It is highly advised that remove the caterpillars by hand.

» Orthene, Sevin, the pyrethroids Talstar and Scimitar, and the biological control agent *Bacillus thuringiensis* are all examples of effective insecticides.

» Caterpillars are parasitized by sarcophagid flies and the braconid wasp *Cardiochiles nigriceps*.

» Many larvae are also killed by polistes wasps.

Mealy bug, *Spilococcus geraniae*

Damage symptoms

» Because they stick to leaves and young stems and feed on the sap, mealybugs can stunt a plant's development.

» Mealy bugs are tiny, creepy-crawly insects that exude a cottony mass to protect their bodies.

» Similarly, they are sucking insects that thrive in warm, moist environments where the tissues are pliable.

» They look like little tufts of cotton and tend to cluster towards the point where the stem meets the leaf.

» Mealy bugs feed off plant liquids and can transfer disease to other plants.

» They eat all a plant has to offer, although they're especially fond of tender, young shoots.

- Infestation symptoms can include leaf yellowing and leaf drop.
- When mealy bugs feed on plants, they secrete a honeydew-like liquid all over the foliage, which then attracts and feeds black sooty mould, giving the plant the appearance of being coated in chimney soot.

Favorable conditions

- Conditions ideal for the development of mealy bugs are those of high heat and low humidity.

Management

- A cotton swab dipped in alcohol will effectively eliminate them.
- Insecticidal soap (a powerful jet of sprayed soapy water) or horticultural oils applied multiple times weekly offer some protection.
- Apply Confidor, White Oil, or Insecticidal Soap spray.
- Try releasing some parasitic wasps or the mealy bug destroyer (*Cryptolaemus montrouzieri*).

DISEASES

Botrytis blight, *Botrytis cinerea*

Symptoms

- The grey, fuzzy growth characteristic of the fungus is most easily identified under wet, humid conditions, when it has had a chance to spread via affected plant tissues and debris.
- Leaves develop a variety of abnormalities, from pinpoint lesions to massive die-offs marked by concentric rings
- Occasionally, leaves will acquire lesions that are V-shaped.
- Dead flower petals that land on leaves often cause lesions.
- Stem lesions, ranging in colour from white to dark brown, first show up at the plants' bases.
- Eventually, these lesions will ruin the entire stem.
- There is a risk of brown lesions appearing on the plants once cuttings have been made.
- Premature browning and falling of flowers can occur when relative humidity and moisture levels are too high.

- A grey, fuzzy mass of the fungus covers the wilting blossoms.
- When geranium flowers become *botrytis*, they lose a lot of their appeal and value.

Favorable conditions

- It has been found that the fungus' lesion size and sporulation rate both rise between 25 and 30 °C, where the disease is most prevalent.
- When relative humidity is at or above 85% and air movement is minimal or nonexistent, the fungus becomes extremely active and reproduces quickly.

Survival

- Conidia, mycelium, or sclerotia of the fungus can overwinter on either living or dead plant matter, allowing it to thrive in the field throughout the year.
- Infection spreads through leftover plant debris, diseased plant components, and contaminated soil.

Management

- Utilization of virus-free vegetation.
- Creating a soil with good drainage by the addition of soil amendments.
- Not use overhead sprinklers and water before 11 a.m.
- Cleaning up crop trash and removing infected plants or plant pieces.
- Quicker leaf drying can be achieved by increasing airflow.
- Iprodione, methylthiophanate, mancozeb, chlorothalonil, or azoxystrobin can be sprayed.
- Some organisms, such as *Trichoderma harzianum* T-22 and *Streptomyces* sp., are utilised as biocontrol agents.

Black leg/Root rot, *Pythium* sp.

Symptoms

- The fungal infection begins at the root tips and spreads upward via the plant's vascular system to the plant's crown.
- Unfortunately, by the time the symptoms manifest, it is usually too late to save the plant because the healthy roots have already died.
- When a plant's roots or stems rot, the plant's upper parts wilt, stunt, and become yellow.

- Since the fungus attack and girdle the plant, a brown to black lesion is often visible at the soil line.
- The bottoms of cuttings rot first, turning a brownish-black colour from water damage.
- Coal black rot travels quickly up the stem, killing the cutting between 7.5 and 10.0 centimetres from the base.
- Infected plants will develop a webby growth that is greyish or off-white in colour when exposed to high humidity.
- When plants are girdled, they die.

Favorable conditions

- Roots can become infected with pythium when the media is very damp and humid.

Management

- Including additives to the soil to enhance water flow.
- Utilization of virus-free vegetation.
- One should stay away from both pools of water and excessive amounts of nitrogen fertiliser.
- Infected plants can be removed from the environment to lower the spread of disease.
- Effective water management is crucial for preventing root rots.
- Use a spray containing Metalaxyl, Mefonoxam, Fosetyl-AL, Etridiazole, or Etridiazole plus Methyl Thiophanate.
- Make use of biocontrol agents such as *Trichoderma harzianum* T-22 and *Streptomyces* sp.

Leaf rust, *Puccinia pelargonii-zonalis*

Symptoms

- Symptoms of leaf rust include yellow dots on the leaf's upper surface and pustules of a rusty tint forming in the spots.
- Once the pustules open, spores are released.
- The spores that develop from a pustule frequently do so in a ring.
- When the disease is bad, the leaves turn yellow and fall off.

- The stems, petioles, and stipules all have pustules that are filled with rust-colored spores.
- The disease is more prevalent in greenhouse production, but it could have significant ecological consequences.

Favorable conditions

- Plant diseases are most common when temperatures and humidity are low.

Survival

- As the disease spreads from plant to plant, the following harvest is doomed because of the sick leaves and other plant parts.

Management

- Weeding and picking up dead leaves and other plant matter are both excellent hygiene practises.
- To improve soil drainage, incorporate organic amendments.
- It's important to use plants that are resistant to diseases.
- Don't water the lawn from above.
- Clean up any infected plants.
- Triadimefon, Mancozeb, Thiophanate Methyl with Mancozeb, Chlorothalonil, Oxycarboxin, Azoxystrobin, Fenhexamid, and Triforine can all be used as soil drenches to kill off any fungi that may be present.

Powdery mildew, *Erysiphe communis*

Symptoms

- The leaves, stems, and flowers can all be covered in a white or grey powdery growth.
- Affect infected foliage by making it look distorted and discoloured before it prematurely withers and falls off.
- Since powdery mildews can only live in living plant tissue, they have adapted to avoid destroying their hosts.

Favorable conditions

- When humidity levels are high, the disease spreads rapidly.

Management

- Identify and remove any rogue sick plants or diseased branches.
- Don't cram too many plants into a small space and make sure there's adequate ventilation.
- Reduce the amount of humidity in the air.
- Eliminate weed hosts and volunteer plants by giving the greenhouse a thorough cleaning in between planting seasons.
- Ornamental plants can be treated with fungicides like Compass, Cygnus, First Step, Pipron, Rubigan, or Zyban, among others.

Leaf spot/blight, *Alterneria alternata*

Symptoms

- Brown necrotic spots form at the leaf's margin first, and they spread inside toward the midrib, eventually causing the entire leaf to die.
- Symptoms manifest initially as tiny, elevated, water-soaked patches on the lower leaf surface.
- The blotches turn brown and sunken very soon.
- Spots caused by *Alternaria* can be easily mistaken for those caused by Bacterial Blight, however the latter are often smaller.
- When humidity levels are high, the Alternaria patches become blanketed in a layer of dark brown, fuzzy spore masses.

Favorable conditions

- *Alternaria* thrives in temperatures between 25 and 28 °C, with rain or heavy dew being ideal.

Survival and spread

- Plant debris that has been infected with *Alternaria* sp. and is left on the ground or in the compost pile, as well as seeds that have been infected by this fungus, can serve as a reservoir for the fungus to spread to subsequent crops.
- Both wind and raindrop splashes can spread spores to neighbouring plants.

Management

- It is important to pick up and dispose of any plant litter or plant debris that may have been contaminated with diseased leaves.

» It's best to water before 11 a.m. so that the leaves don't sit in moist soil all day.

» It is highly advised that annuals be thinned out and that perennials be pruned deliberately to maximise airflow and light penetration.

» 0.3% Zineb or 0.1% Benomyl is sprayed on the crop every 12-15 days.

» The fungicides Iprodione, Thiophanate Methyl plus Iprodione, Mancozeb, Chlorothalonil, Thiophanate methyl plus Chlorothalonil, Azoxystrobin, and Fludiononil are used to combat Alternaia leaf spot disease.

Bacterial blight, *Xanthomonas campestris* pv. *pelargonii*

Symptoms

» This bacterium enters the plant through wounds and other holes, and it multiplies and spreads throughout the plant's vascular system.

» When bacteria multiply, they block the plant's vascular system and prevent water from reaching the plant's leaves, causing the plant to wilt.

» The bacteria can infect the entire crop if the sick plant is left unchecked for even a short period of time.

» A symptom of a Xanthomonas infection is V-shaped lesions on the leaves of the affected plant.

» Small, distinct, water-soaked or brown and V-shaped dots first appear on the underside of affected leaves.

» There is a definite darkening of the veins and a wilting of the leaf edges in certain leaves.

» Any time an disease spreads throughout the plant, the whole thing droops and the individual branches die off.

» The stems turn black and dry up as they rot.

» Diseased seedlings can't get their bearings and eventually decay from the bottom up.

» The stems turn a flat, dark brown colour. When plants are infected systemically, they wilt and die.

Favorable conditions

» Above 70 °F and below 81 °F is ideal for the progression of disease.

Survival and spread

- For up to a year, the bacterium can live in decomposing plant matter in soil.
- Infected plants can transmit the germs to healthy plants via pruning shears.
- It is possible for bacteria to be transferred from one plant to another by water splashes and from one cutting to another through contaminated growth medium.

Management

- Scavenge for trash and clean up plant debris.
- Create well-drained soil, use soil amendments.
- It's important to use plants that are resistant to diseases.
- Water early in the day, preferably before 11 a.m., and avoid watering from above.
- Never plant a perennial geranium (*Geranium* spp.), such as a *geranium*, next to a *Pelargonium*.
- Diseased plants should be removed immediately.
- Plants strategically placed to improve ventilation.
- Don't touch plants unless absolutely necessary, and make sure to wash your hands often.
- Geranium's bacterial blight can be treated with copper sulphate pentahydrate.

Bacterial leaf spot, *Pseudomonas syringae*

Symptoms

- Signs on leaves range from pinpoint lesions to massive die-offs, sometimes arranged in concentric circles.
- Lesions in the shape of a V may also appear.
- Stem lesions, ranging in colour from white to dark brown, first show up at the plants' bases.
- Eventually, these lesions will ruin the entire stem.
- There is a risk that flowers could turn brown and fall off early.

Favorable conditions

- High relative humidity, prolonged leaf wetness, and elevated nitrogen levels all contribute to the spread of disease.

Spread

- Splashing water around is a great way to transmit bacteria.

Management

- Maintain proper field hygiene.
- Create a soil that drains well with the help of soil amendments.
- It's important to use plants that are resistant to diseases.
- Irrigating early in the day or using a sub-irrigation system will minimise splashing and shorten the amount of time the leaves are wet.
- Water early in the day, preferably before 11 a.m., and avoid watering from above.
- Copper fungicide spray.

References

Abdel-Rahman, T. F., Abdel-Wahed, G. A., & Hassanin, M. M. (2023). Characterization and Control of Root Rot and Wilt Fungi of Geranium (Pelargonium graveolens L.) with Special Reference to First Record for Fusarium proliferatum as Wilt Pathogen on Geranium in Egypt. Egyptian Journal of Phytopathology, 51(1), 122-141.

Prasad, D., Singh, K. P., & Kumar, J. (2008). Root rot and wilt complex of rose-scented Geranium (Pelargonium graveolens) in Uttarakhand. Indian Phytopathology, 61(3), 365.

Liang, Z., Lai, K., Zhu, Y., Chen, W., Fang, F., & Ye, Y. (2010). Investigation on species of diseases and pests in Pelargonium graveolens L. Her and control technique. Plant Diseases and Pests, 1(3), 51-54.

Tucker, A. O. (2004). Geranium and pelargonium: the genera Geranium and Pelargonium. Journal of Essential Oil Research: JEOR, 16(3), 160.

FSC
www.fsc.org
MIX
Papier | Fördert
gute Waldnutzung
FSC® C083411

Kolibri 360 GmbH
Ferdinand-Jühlke-Straße 7
99095 Erfurt, Deutschland
produktsicherheit@kolibri360.de